ZEITSCHRIFT FÜR ANGEWANDTE MATHEMATIK UND MECHANIK

INGENIEURWISSENSCHAFTLICHE FORSCHUNGSARBEITEN

BAND 54 | 1974 | HEFT 2

AUS DEM INHALT:

AKADEMIE-VERLAG GMBH · BERLIN

ZAMM | Bd. 54 | Nr. 2 | S. 73—132 | Berlin, Februar 1974

M 18,—

INHALT

Wir bitten, **Manuskriptsendungen** an folgende Anschrift zu richten:

Zeitschrift für Angewandte Mathematik und Mechanik
z. Hd. Herrn Prof. Dr. Günter Schmidt
Zentralinstitut für Mathematik und Mechanik
an der Akademie der Wissenschaften der DDR
DDR 108 Berlin, Mohrenstraße 39.

Zu den Arbeiten, die als Hauptaufsätze bestimmt sind, erbitten wir auf gesondertem Blatt eine kurze Zusammenfassung des Inhalts, nach Möglichkeit in deutscher, englischer und russischer Sprache; falls die Übersetzungen nicht geliefert werden können, ist wenigstens die Angabe spezieller Fachausdrücke in den verschiedenen Sprachen erwünscht. Ausführliche Hinweise für die Autoren, um deren strikte Berücksichtigung gebeten wird, finden sich im Anschluß an das Inhaltsverzeichnis des Jahrganges **53** (1973).

Zuschriften, welche Manuskripte und Arbeiten betreffen, deren Eingang von dem jetzigen Herausgeber Prof. Dr. H. Heinrich bestätigt worden ist, sind weiterhin an ihn unter der folgenden Anschrift zu richten:

Zeitschrift für Angewandte Mathematik und Mechanik
z. Hd. Herrn Prof. Dr.-Ing. Dr. h. c. Helmut Heinrich
DDR 8027 Dresden, Friedrich-Hegel-Str. 31.

Die Autoren erhalten von den Hauptaufsätzen 75, von den Kleinen Mitteilungen 25 Sonderdrucke ohne Berechnung, darüber hinaus bis zu 250 Sonderdrucke gegen Berechnung.

ZAMM 54, 73 – 82 (1974)

A. N. SHERBOURNE / H. M. HAYDL

Plastic Analysis of Shallow Spherical Shells under Combined Loading at Moderately Large Deflections

Für das im Titel genannte Problem werden die Gleichungen aufgestellt und eine numerische Approximation und eine genauere Lösung gegeben. Der Einfluß einer Querkraft auf das Fließen der Schalen wird berücksichtigt. Für zwei Schalen, für die das Fließgesetz von Huber- v. Mises gilt, werden die Wechselwirkungsflächen für die Traglast, die Spannungsverteilungen und die Verschiebungen aufgezeigt. Die Ergebnisse werden diskutiert und die Erweiterung der numerischen Methode auf kompliziertere Probleme angedeutet.

Consistent equations, numerical approximate and improved solutions are given for the title problem. The influence of transverse shear on the plastification of the shells is included in the analysis. Load interaction surfaces at collapse, stress fields and displacement fields are shown for two shells, whose material yields according to the Huber- v. Mises criterion. The results are discussed and the extension of the numerical approach to more complex problems is indicated.

Для определённой заглавием задачи выводятся уравнения и определяются приблизительное численное и более точное решения. Учитывается влияние поперечной силы на пластическую деформацию оболочек. Для двух оболочек, подчиняющихся закону пластической деформации Губера -фон Мизеса, определены поверхности противоположного действия для несущей нагруски, распределения напряжений и смещений. Результаты подвергаются дискуссии. Дано указание о возможности распространения численного метода на более сложные задачи.

Symbols

A, B, K, K_1	constants	n_0	non-dimensional edge load, N_r/N_0
c, h, α	non-dimensional shell parameters	p, P	lateral and non-dimensional lateral pressure
H, R, R_1	shell dimensions (Fig. 1)	P^*	'punching shear' load
$\dot{k}_i, \dot{\varepsilon}_{ij}, \dot{\varepsilon}^0_{ij}$	curvature rates and strain rates	r, θ, z	coordinate directions
$k_i, \varepsilon_{ij}, \varepsilon^0_{ij}$	curvatures and strains	σ_{ii}, τ_{rz}	stresses
N_i, M_i, Q	stress resultants	u, w, β	displacements
n_i, m_i, q	non-dimensional stress resultants	U, W	non-dimensional displaecments
$\bar{M}_r, \bar{N}_r, \bar{Q}$	edge loads	$\dot{u}, \dot{w}, \dot{\beta}$	velocities
M_0	fully plastic moment	$\dot{U}, \dot{W}$	non-dimensional velocities
N_0	fully plastic membrane force	ξ	non-dimensional radial coordinate

1. Introduction

In classical limit analysis [1] it is assumed that a structural member, of a rigid-plastic material, remains rigid under a monotonically increasing load until the limit load is reached. It is further assumed that, from this point on, uncontained plastic flow occurs without any changes in the limit load. In reality the load must be increased or decreased during quasi-static plastic flow; these changes in load can be explained as being due to changes in geometry. When changes in geometry are considered in the plastic analysis one is led to the problem of post-yielding and the member may exhibit "geometrical strengthening" or "geometrical weakening". A theory containing geometric effect will therefore give a more accurate description of the real phenomenon.

Limit analyses (small deformation theory) of shallow shells have been considered by several authors [2]–[11]. All these solutions are based on the TRESCA yield condition and neglect the influence of shear on yielding. Post yielding (large deformation theory) of shallow shells has received little attention in the literature. DERKACH and SHEVCHENKO [12] have given an example of a shallow spherical shell which geometrically strengthens during plastic flow, i.e. requires an increase in load with increasing deflections. DUSZEK [13] has given an approximate solution for a similar problem and SHABLII [14] has examined a problem of geometrical strengthening and weakening of a shallow shell. The above non-linear solutions are based on the TRESCA yield condition and neglect transverse shear effects on the plastification of the shell. For the HUBER-v.-MISES yield condition no solution is known for plastic shallow shells at large deflections.

In this paper the collapse and post-collapse of shallow spherical shells at moderately large deflections (i.e. lateral deflections of the order of the shell thickness) are considered. It is assumed that the shells are made of rigid-plastic material which yields according to the HUBER-v.-MISES criterion; the effects of geometry changes on the limit load is investigated. Transverse shear effects on the plastification of the shells are accounted for, thus making the analysis applicable to thin and moderately thick shallow shells. The loading considered is a combined uniformly distributed lateral pressure and an evenly distributed edge compression. The combined external loads are taken to be of the proportional type. The outer contour of the shells is assumed to be simply or hinge supported.

The proposed method of solution is a variation of the iterative scheme used by SHERBOURNE [15] in the investigation of post-buckling of circular plates. In the following we show a simple approximate solution and an improved solution. Results are given for some examples of shallow shells.

2. Shallow Spherical Shells

2.1. Assumptions of shallow shell theory

The analysis is restricted to an investigation of "shallow" spherical shells. These shells, whose geometry is shown in Fig. 1, must fulfil the following geometric conditions.

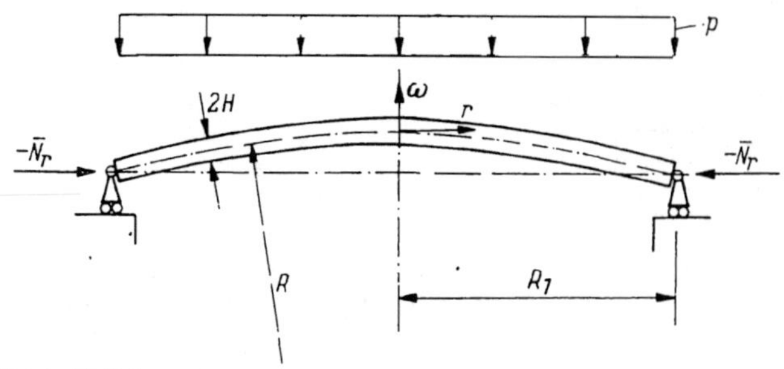

Fig. 1. Shell Geometry

a) Reissner's [16] definition of "shallow" applies, namely

$$\frac{R_1}{R} \leqq 0.5 \,. \tag{1}$$

b) The first principal quadratic form of the shell reference surface is taken as

$$ds^2 = dr^2 + r^2 \, d\theta^2 \,. \tag{2}$$

c) The geometric relations, connecting the coefficients of the first and second principal quadratic forms, of Codazzi [17] are satisfied exactly, whereas the condition of Gauss [17] is satisfied only approximately if we require that

$$R \gg R_1 \,. \tag{3}$$

2.2. Strain-displacement relations

Using the assumption made for shallow shells, eq. (2), we assume the displacements of the shell as

$$U(r, z) = u(r) + z\,\beta(r) \,, \tag{4}$$

$$W(r, z) = w(r) \,. \tag{5}$$

The strain components of a point in the shallow shell undergoing large deformations are [18]

$$\varepsilon_{rr} = \frac{\partial U}{\partial r} + \frac{W}{R} + \frac{1}{2}\left[\left(\frac{1}{2}\frac{\partial U}{\partial r} + \frac{W}{R}\right)^2 + \left(\frac{\partial W}{\partial r} - \frac{U}{R}\right)^2\right], \tag{6}$$

$$\varepsilon_{\theta\theta} = \frac{U}{r} + \frac{W}{R} + \frac{1}{2}\left[\frac{U}{r} + \frac{W}{R}\right]^2, \tag{7}$$

$$\varepsilon_{rz} = \frac{\partial U}{\partial z} + \frac{\partial W}{\partial r} - \frac{U}{R} + \frac{\partial U}{\partial z}\left(\frac{\partial U}{\partial r} + \frac{W}{R}\right). \tag{8}$$

If the displacement field equations (4) and (5) are substituted into (6) to (8) the strain-displacement relations for large deformations are obtained. The resulting equations are rather complicated and unsuitable for engineering estimates. Some simplifications can be made to make the equations useful for a study of "moderately" large deformations. For this purpose the terms are ordered according to their magnitude as shown in Table 1 ($\bar{\varepsilon} \ll 1$).

Table 1

Order of Magnitude	1	$\varepsilon^{1/2}$	$\bar{\varepsilon}$	$\bar{\varepsilon}^2$	$\bar{\varepsilon}^\gamma$, $\gamma > 2$
Classified terms	$\frac{w}{2H}$	c	w, β	h, u, u'	
		α	w', β'	$\frac{w}{R}, \frac{u}{R}, \frac{\beta}{R}$	

If terms of the order $\bar{\varepsilon}^\gamma$, $\gamma > 2$, are neglected in the strain-displacement relations the following implicit assumptions have been made:

a) the strains are small in comparison to unity and the inplane displacements are small when compared to R,
b) the deflections are "moderately" large, but small when compared to the radius R,
c) the rotations of the cross sections are larger, and
d) the slope may be large.

We also assume that the shear strain ε_{rz} does not vary through the shell wall thickness and that for shallow shells, according to Reissner [16], the influence of the inplane displacement u on the shear strain ε_{rz} is negligible. As a consequence of the latter assumption we neglect the influence of the shear force on the inplane equilibrium equation.

Including all of the above assumptions results in the strain-displacement relations for shallow spherical shells at moderately large deflections. These relations are

$$\varepsilon_{rr} = \varepsilon^0_{rr} + z\,k_r \,, \qquad \varepsilon_{\theta\theta} = \varepsilon^0_{\theta\theta} + z\,k_\theta \,, \qquad \varepsilon_{rz} = \varepsilon^0_{rz} \,, \tag{9), (10), (11}$$

where

$$\varepsilon^0_{rr} = u' + \frac{w}{R} + \frac{1}{2}(w')^2 , \qquad \varepsilon^0_{\theta\theta} = \frac{u}{r} + \frac{w}{R}, \qquad \varepsilon^0_{rz} = w' + \beta ,$$

$$k_r = \beta' , \qquad k_\theta = \frac{1}{r}\beta .$$

In the above primes denote differentiation with respect to the coordinate direction r.

To obtain the strain rate-velocity relations we differentiate (9) to (11) with respect to a time like parameter (indicated by a dot). The relations obtained are

$$\dot\varepsilon_{rr} = \dot\varepsilon^0_{rr} + z\,\dot k_r , \qquad \dot\varepsilon_{\theta\theta} = \dot\varepsilon^0_{\theta\theta} + z\,\dot k_\theta , \qquad \dot\varepsilon_{rz} = \dot\varepsilon^0_{rz} \qquad (12)\ (13)\ (14)$$

where

$$\dot\varepsilon^0_{rr} = \dot u' + \frac{\dot w}{R} + w'\,\dot w' , \qquad \dot\varepsilon^0_{\theta\theta} = \frac{\dot u}{r} + \frac{\dot w}{R}, \qquad \dot\varepsilon^0_{rz} = \dot w' + \dot\beta ,$$

$$\dot k_r = \dot\beta' , \qquad \dot k_\theta = \frac{1}{r}\dot\beta .$$

2.3. Equilibrium equations

Since theorems of limit analysis are based on the principle of virtual power, this principle is used to drive the equilibrium equations and the boundary conditions.

This principle states that

$$\dot V - W_s - W_e = 0 . \qquad (15)$$

Here the strain energy is

$$V = \int_{\text{Vol.}} (\sigma_{rr}\,\dot\varepsilon_{rr} + \sigma_{\theta\theta}\,\dot\varepsilon_{\theta\theta} + \tau_{rz}\,\dot\varepsilon_{rz})\, r\, dr\, d\theta\, dz , \qquad (16)$$

the work of the surface forces is

$$W_s = - \int_{\text{Surf.}} p\,\dot w\, r\, dr\, d\theta \qquad (17)$$

and the work of the edge forces is (on r = constant)

$$W_e = \int_\theta (\bar N_r\,\dot u + \bar Q\,\dot w + \bar M_r\,\dot\beta)\, r\, d\theta . \qquad (18)$$

The stress resultants are defined as

$$N_i = \int \sigma_{ii}\, dz , \qquad M_i = \int \sigma_{ii}\, z\, dz , \qquad Q = \int \tau_{rz}\, dz , \qquad i = r, \theta . \qquad (19)$$

The positive direction of the stress resultants is shown in Fig. 2.

Upon substituting (12) to (14) and (16) to (19) into (15) and performing the indicated integrations, the following equilibrium equations and boundary conditions are obtained

$$-(r\,N_r)' + N_\theta = 0 , \qquad (20)$$

$$\frac{r}{R}(N_r + N_\theta) - (r\,N_r\,w')' - (r\,Q)' + p\,r = 0 , \qquad (21)$$

$$-(r\,M_r)' + M_\theta + r\,Q = 0 \qquad (22)$$

the boundary conditions on r = constant are

$$\dot u\, r\,(N_r - \bar N_r) = 0 , \qquad \dot w\, r\,(N_r w' + Q - \bar Q) = 0 ,$$

$$\dot\beta\, r\,(M_r - \bar M_r) = 0 . \qquad (23)$$

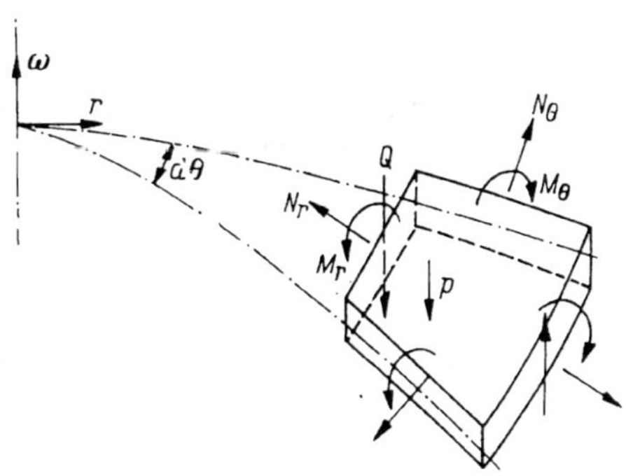

Fig. 2. Positive stress resultants

2.4. Yield surface and flow laws

The yield surface used in this analysis is an approximation to the yield surface of Ilyushin [19]. This approximate yield surface has first been proposed by Shapiro [20] and has also been derived by Schroeder [21]. Without the inclusion of transverse shear terms the accuracy of this surface has been examined by Robinson [22], who estimates that collapse load obtained with this surface are in the range $0.955\,P_0 \leq P \leq 1.155\,P_0$. Here P_0 is the collapse load based on the exact Ilyushin yield surface. Haydl and Sherbourne [23] have suggested that inclusion of transverse shear does not alter significantly these error bounds on the collapse load P. This yield surface is given by

$$\frac{1}{N_0^2}(N_r^2 + N_\theta^2 - N_r N_\theta + 3\,Q^2) + \frac{1}{M_0^2}(M_r^2 + M_\theta^2 - M_r M_\theta) = 1 \qquad (24)$$

or in general form

$$f(N_i, M_i, Q) = \text{constant}\,. \tag{25}$$

The equations describing the motion of the yield mechanism are furnished by flow theory [24]. Flow theory assumes that the yield surface is the potential for the corresponding strain rates, i.e.

$$\dot{\varepsilon}^0_{ii} = \dot{\lambda}\frac{\partial f}{\partial N_i}, \qquad \dot{k}_i = \dot{\lambda}\frac{\partial f}{\partial M_i}, \qquad \dot{\varepsilon}^0_{rz} = \dot{\lambda}\frac{\partial f}{\partial Q}, \qquad i = r, \theta\,. \tag{26}$$

Here $\dot{\lambda}$ is a non-negative flow parameter. Eq. (26) are the equations for the incremental strain theory. If the strain rates are replaced by the strains we obtain the equations of the total strain or deformation theory.

HILL [25], DRUCKER [26] and ILYUSHIN [27] have shown that the predictions of the deformation theory agree with the incremental strain theory for the restrictive condition that principal directions of stress remain constant and that the stress ratios are held constant. This is the condition for proportional loading of the volume element. BUDIANSKY [28] has shown that deformation theory can be used for other loading paths than proportional loading without violating the general requirements and soundness of a plasticity theory.

Since the governing equations of our problem, when the incremental strain theory is considered, are extremely complex and difficult to solve, we shall make use of the deformation theory.

2.5. Non-dimensional equations

The following non-dimensional terms are introduced at this point:

$$\left.\begin{array}{llll} n_r = \dfrac{N_r}{N_0}, & n_\theta = \dfrac{N_\theta}{N_0}, & m_r = \dfrac{M_r}{M_0}, & m_\theta = \dfrac{M_\theta}{M_0}, \\[2ex] q = \sqrt{3}\,\dfrac{Q}{N_0}, & N_0 = 2\,H\,\sigma_0\,, & M_0 = \sigma_0\,H^2\,, & \xi = \dfrac{r}{R}, \\[2ex] P = p\dfrac{R}{N_0}, & \alpha = \dfrac{2\,H}{2\,R_1}, & \dot{W} = \dfrac{\dot{w}}{R}, & W = \dfrac{w}{R}; \\[2ex] \dot{U} = \dfrac{\dot{u}}{R}, & c = \dfrac{R_1}{R}, & h = \alpha\,\dfrac{c}{2}. & \end{array}\right\} \tag{27}$$

Using this notation the equilibrium equations (20) to (22) are written in non-dimensional form as

$$(\xi\,n_r)' - n_\theta = 0\,, \tag{28}$$

$$n_r\,(\xi - W') - \frac{1}{\sqrt{3}}\,q + \frac{1}{2}\,\xi\,P = 0\,, \tag{29}$$

$$(\xi\,m_r)' - m_\theta - \frac{1}{\sqrt{3}\,h}\,\xi\,q = 0\,. \tag{30}$$

Note that in (29) n_θ has been replaced by its value as obtained from (28).

The yield surface (24) becomes

$$n_r^2 + n_\theta^2 - n_r\,n_\theta + m_r^2 + m_\theta^2 - m_r\,m_\theta + q^2 - 1 = 0\,. \tag{31}$$

The combination of (9) to (11), (24) and (26) leads to the displacement field equations

$$\frac{\lambda}{N_0}(2\,n_r - n_\theta) = U' + W + \frac{1}{2}(W')^2\,, \tag{32}$$

$$\frac{\lambda}{N_0}(2\,n_\theta - n_r) = \frac{U}{\xi} + W\,, \qquad \frac{\lambda}{N_0}(2\,m_r - m_\theta) = h\,\beta'\,, \tag{33), (34}$$

$$\frac{\lambda}{N_0}(2\,m_\theta - m_r) = \frac{h}{\xi}\,\beta\,, \qquad \frac{\lambda}{N_0}\,2\,\sqrt{3}\,q = W' + \beta\,. \tag{35), (36}$$

Similarly, the velocity field equations are obtained as

$$\frac{\dot{\lambda}}{N_0}(2\,n_r - n_\theta) = \dot{U}' + \dot{W} + W'\,\dot{W}'\,, \tag{37}$$

$$\frac{\dot{\lambda}}{N_0}(2\,n_\theta - n_r) = \frac{1}{\xi}\,\dot{U} + \dot{W}\,, \qquad \frac{\dot{\lambda}}{N_0}(2\,m_r - m_\theta) = h\,\dot{\beta}' \tag{38), (39}$$

$$\frac{\dot{\lambda}}{N_0}(2\,m_\theta - m_r) = \frac{h}{\xi}\,\dot{\beta}\,, \qquad \frac{\dot{\lambda}}{N_0}\,2\,\sqrt{3}\,q = \dot{W}' + \dot{\beta}\,. \tag{40), (41}$$

In the above equations a prime denotes differentiation with respect to the variable ξ. The boundary condition (23) are, in non-dimensional form, for our problem at the shell edge $\xi = c$:

$$n_r = n_0\,, \qquad \dot{W}\ (\text{or } W) = 0\,, \qquad m_r = 0\,, \tag{42), (43), (44}$$

at the shell centre $\xi = 0$:

$$U\ (\text{or } \dot{U}) = 0\,, \qquad \beta\ (\text{or } \dot{\beta}) = 0\,, \qquad W\ (\text{or } \dot{W}) = W_0\ (\text{or } \dot{W}_0)\,, \tag{45), (46}$$

and from symmetry considerations at $\xi = 0$:

$$n_r = n_\theta\,, \qquad q = 0\,, \qquad m_r = m_\theta\,. \tag{47}$$

For the incremental theory we specific the initial conditions at time $t = 0$:

$$U = W = \beta = 0\,. \tag{48}$$

For the incremental theory the complete set of governing equations is given by the 9 equations (28) to (31) and (37) to (41). A solution of this system appears to be extremely difficult and will not be attempted here.

The deformation theory equations are (28) to (36). Although this system of 9 equations is non-linear and the stress and velocity yields are coupled a numerical solution is possible and will be outlined subsequently.

3. Approximate Solution for Shallow Spherical Shells

According to limit analysis theorems [1] a lower bound on the exact collapse load is obtained if equilibrium is satisfied and the yield condition is not violated. In this section an approximate (lower bound) solution will be outlined. The solution is based on an assumption first suggested by MIKELADZE [2]. This assumption takes the form

$$n_r = n_\theta = n_0 = \text{constant} \tag{49}$$

throughout the shell.

3.1. Stress field and collapse loads

With the assumption (49) the equilibrium equation (28) is identically satisfied and eqs. (29) to (31) become

$$n_0\,(\xi - W') - \frac{1}{\sqrt{3}}\,q + \frac{1}{2}\,\xi\,P = 0\,, \tag{50}$$

$$(m_r\,\xi)' - m_\theta - \frac{1}{\sqrt{3}\,h}\,\xi\,q = 0\,, \tag{51}$$

$$n_0^2 + m_r^2 + m_\theta^2 - m_r\,m_\theta + q^2 - 1 = 0\,. \tag{52}$$

The approximate solution is given by solving the above 3 equations subject to the appropriate boundary conditions. The problem becomes statically determinate if we assume the deflected shape as

$$W = K\,(\xi^2 - c^2)\,. \tag{53}$$

With this value of W eq. (51) can be solved for

$$q = \frac{\sqrt{3}}{2}\,\xi\,P + \sqrt{3}\,\xi\,n_0\,(1 - 2\,K)\,. \tag{54}$$

Combining (52) and (53) results in

$$m_r' = \frac{1}{\sqrt{3}\,h}\,q - \frac{m_r}{2\,\xi} - \frac{1}{\xi}\left(-\frac{3}{4}\,m_r^2 - q^2 + 1 - n_0^2\right)^{1/2}. \tag{55}$$

For assumed value of c, h, n_0, K and P the stress field can be determined from (51), (54) and (55).

The equations are examined to see if a continuous plastic stress yield exists. For this purpose the yield surface equation (52) is written as

$$m_\theta = \frac{1}{2}\,m_r \pm \left(-\frac{3}{4}\,m_r^2 - q^2 + 1 - n_0^2\right)^{1/2}. \tag{56}$$

This equation describes two plastic regimes which are separated by the line

$$m_\theta = \frac{1}{2}\,m_r\,. \tag{57}$$

where

$$m_r = \pm\frac{2}{\sqrt{3}}\,(1 - q^2 - n_0^2)^{1/2}\,.$$

From considerations of the shell geometry and the boundary conditions it is evident that the plastic stress yield is continuous and that only the plastic regime

$$m_\theta = \frac{1}{2} m_r - \left(- \frac{3}{4} m_r^2 - q^2 + 1 - n_0^2 \right)^{1/2}$$

applies to our problem.

Consider the case where collapse occurs by pure shear ($q = 1$) at the hinged support. This is the collapse mode called "punching shear". From (54) we solve for

$$P^* = \frac{2}{\sqrt{3}\, c} - 2\, n_0 (1 - 2\, K)\,. \tag{58}$$

For moderately large deflections this collapse load interaction surface is well above the interaction surface for the shells examined. We conclude therefore that collapse by "punching shear" does not occur. We examine the pure membrane solution of the lower bound solution. Assume that

$$n_r = n_\theta = -1 \tag{59}$$

throughout the shell, and that $m_r = m_\theta = 0$. From (51) it follows that $q = 0$, i.e. the pure membrane solution is the same if we consider shear or neglect shear in the yield surface equation. Equation (50) can be integrated for W and, by using the proper boundary conditions, results in

$$W = \frac{1}{2} \left(1 - \frac{P}{2} \right) (\xi^2 - c^2)\,. \tag{60}$$

The relation connecting the collapse load P with the central deflection is obtained from this equation as

$$P = 2 + \frac{4\, W_0}{c^2}. \tag{61}$$

It can easily be verified that the deflected shape equation (60) does not satisfy the displacement field equation, and the pure membrane solution is therefore only a lower bound.

3.2. Displacement field

The velocity field equation (37) to (41) can in general not be satisfied with the stress field obtained in the approximate solution. There exists however a special case where the approximate solution coincides with the exact solution of the incremental theory. This is the case when $n_0 = 0$ and it can easily be shown that (37) and (38) admit the solution

$$W = \frac{1}{2} (\xi^2 - c^2) \tag{62}$$

and

$$U = -\frac{1}{2} \xi (\xi^2 - c^2) \tag{63}$$

the rotation of the normal is obtained from (39) and (40) as

$$\beta' = \frac{\beta}{\xi} \frac{2\, m_r - m_\theta}{2\, m_\theta - m_r} \tag{64}$$

and it follows that (40) and (41) are satisfied within a negligible small error (0.3$^0/_0$) when compared to the subsequent results of the improved solution. The central deflection is

$$W_0 = -\frac{1}{2} c^2\,. \tag{65}$$

This central deflection makes the shell essentially flat. Since all equations are satisfied (62) is also the exact solution.

3.3. Numerical scheme

The numerical integration for the approximate solution is relatively simple and the procedure is as follows:

a) Assume value for the shell parameters c and h, the deflection parameter K and the edge load n_0.
b) Assume a value for the lateral load P.
c) Divide the shell radius into a number of equal intervals (we used 50) and calculate q from (54).
d) Integrate (55) with the standard 4th. order RUNGE-KUTTA [29] method, for m_r.
e) From (51) calculate m_θ.
f) Adjust the value of P and go to c) and repeat procedure until the boundary condition (44) is satisfied.
g) Repeat calculations for various combinations of c, h, K and n_0.

During the integration of (55) it is possible that the radical on the right hand side can take on imaginary values. To avoid this use is made of auxilliary yield surfaces whose equations are

$$m_\theta = \frac{1}{2} m_r - \left(\frac{3}{4} m_r^2 + q^2 - 1 + n_0^2\right)^{1/2} .$$

The choice of these auxilliary yield surfaces is completely arbitrary and merely an aid in the machine calculations. The final results must satisfy the yield surface equation (52).

4. Improved Solution for Shallow Spherical Shells

An improved (exact) solution of the eqs. (28) to (36) for the deformation theory is attempted in this section. The proposed numerical scheme is a variation of the scheme proposed by SHERBOURNE [15] in the solution of the post-buckling of circular plates.

4.1. Summary of equations

For the numerical integration scheme the relevant equations are summarized.

Equilibrium equations (28) and (29) remain unchanged as

$$(\xi\, n_r)' - n_\theta = 0\,, \qquad n_r\,(\xi - W') - \frac{1}{\sqrt{3}} q + \frac{1}{2} \xi\, P = 0\,. \tag{67), (68}$$

From (31) we solve for

$$m_\theta = \frac{m_r}{2} - \left(-\frac{3}{4} m_r^2 - q^2 + 1 - n_r^2 - n_\theta^2 + n_r\, n_\theta\right)^{1/2} . \tag{69}$$

This value of m_θ is inserted into (30) and results in

$$m_r' = \frac{q}{\sqrt{3}\, h} - \frac{m}{2\,\xi} - \frac{1}{\xi}\left(-\frac{3}{4} m_r^2 - q^2 + 1 - n_r^2 - n_\theta^2 + n_r\, n_\theta\right)^{1/2} . \tag{70}$$

Combining (34) and (35) gives the rotation of the normal

$$\beta' = \frac{\beta}{\xi}\left[\frac{2\, m_r - m_\theta}{2\, m_\theta - m_r}\right] \tag{71}$$

and from (36) we solve for

$$W' = \beta\left[\frac{2\sqrt{3}\, q\, h}{\xi\,(2\, m_\theta - m_r)} - 1\right] . \tag{72}$$

The approximate inplane displacement U is obtained from combining (32) and (33). The equation is

$$U' = \frac{U}{\xi} + \frac{3\, h}{\xi} \beta \frac{n_r - n_\theta}{2\, m_\theta - m_r} - \frac{1}{2} (W')^2 . \tag{73}$$

The system of 7 equations (67) to (73) is coupled in stress and displacement fields and contains 8 unknowns, namely n_r, n_θ, q, m_r, m_θ, U, β and W.

The equations are solvable if one of the unknowns is assumed within some parameters, and the remaining unknowns are calculated from the equations. Subsequently the free parameters in the assumed unknown are adjusted until finally all 7 equations and all boundary conditions are satisfied and the resulting collapse load P is maximized.

Since in the above we had reduced the original 9 equations in 9 unknowns to 7 equations in 8 unknowns, the obvious choice for the start of our scheme is either n_r (or n_θ). By assuming e.g. n_r as known the system is reduced to 7 equations in 7 unknowns.

4.2. Numerical scheme

The numerical scheme for the improved solution is outlined below.

a) Assume values for the shell parameters h and c. The deflected shape is initially taken as

$$W = K_1\,(\xi^2 - c^2)\,. \tag{74}$$

b) The starting value for the load P is taken from the approximate solution.

c) The stress resultant profile for n_r is taken to be of similar shape as given by DERKACH and SHEVCHENKO [12]. The profile for n_r which gave the best results is of the simple form

$$n_r = A \sin \frac{\pi\, \xi}{c} + B\,(\xi - c) + n_0\,. \tag{75}$$

Here A and B are arbitrary constants.

d) From (67) the value of n_θ can be found.

e) The shell radius is divided into a number of equal intervals and the shear q is calculated from (68).

f) Equation (70) is integrated for m_r with the 4th order RUNGE-KUTTA [29] method and m_θ is found from (69). So far in the scheme only the stress field which satisfies equilibrium and the yield condition at a certain deflection has been found. The boundary condition (44) is not necessarily satisfied at this point. The next step is the evaluation of the displacement field.

g) With the RUNGE-KUTTA method integrate (71) for the rotation of the normal and (72) for the lateral deflection W.

h) Compare the assumed value of W from (74) with the value W as obtained from (72), average the two values and repeat the calculations from step (e) on. After 8 to 10 iterations the two values for W converge to a common value.

i) Solve (73) for the approximate displacement U.

j) Repeat the calculation for different values of P from step c) on until the boundary condition (44) is satisfied.

k) Repeat the scheme for various combinations of the parameters K_1, h, c, A, B, and n_0 until a maximum P is found.

Note that in the scheme the appropriate boundary conditions are satisfied at each step in the calculation of the displacement field. Although the magnitude of the collapse load P is not very sensitive to small changes in the assumed stress resultant profile for n_r, it is observed that the magnitudes of the displacements are extremely sensitive to these changes.

5. Results and Discussion

Before discussing the particular shells examined, the nature and significance of the "interaction" surfaces shown in Figs. 3 and 7 are pointed out. The improved solution, as in the approximate solution, consists of finding limit loads of initially deformed shallow shells, the feature here being that the assumed initial shape is corrected until a kinematically admissible displacement field is found. The "interaction" surface is then constructed by assembling the solutions, i.e. stacking n_0-P interaction curves along the W_0 axis (Figs. 3 and 7). It is suggested that "interaction" surfaces can be used to predict the post-yield behaviour of the spherical shallow shells. Consider the load paths B in Fig. 3 and 7, which are obtained as the intersections of load paths A (in planes parallel to the n_0-P plane at several values of W_0) with the "interaction" surface. Paths B describe the condition of the shell at different deformation stages, i.e. for quasi-static flow to occur the external loads have to be changed as indicated along paths B. The deformed shapes of the shell during plastic flow are given by the solution. The "interaction" surfaces are therefore not load interaction curves in the usual sense of classical plasticity. Along paths B no unloading of the shell elements occurs, and the surfaces are restricted to proportional loading.

The individual results are discussed next.

Shell 1: In Fig. 3 is shown the interaction surface for a shell of $c = 0.1$, $\alpha = 1/20$. The load interactions n_0-P are plotted as functions of the central deflection W_0. Results are shown for the approximate and the improved solution. For small W_0 the approximate solution gives a reasonably good approximation for the collapse loads. At large W_0 the approximate predictions are quite conservative. The interaction surface, both for the approximate and the improved solution, is double valued for $0 \leqq W \leqq \frac{1}{2} c^2$. At $W_0 = \frac{1}{2} c^2$ the shell is almost flat and for $W_0 \leqq \frac{1}{2} c^2$ the load interaction becomes single valued.

For values of the edge force $n_0 \leq 0.15$ we obtain unloading-loading curves, i.e. the influence of geometry changes results in weakening of the shell and after reaching the flat position the shell strengthens. The approximate solution gives only "weakening" of the shells. At a value of $n_0 > 0.15$ we observe only geometrical weakening with increasing deflections for the improved solution. Fig. 4 shows some typical stress fields for this shell. Note that the edge zone of the shell goes from tension to compression yielding with increasing deflections. In the central zone yielding is due mainly to m_θ. In Fig. 5 are plotted deformed shapes of the shell for several edge loads n_0 at different deflections. Fig. 6 gives the corresponding rotations β and edge displacements U.

All results shown in Figs. 4—6 are for the improved solution.

Shell 2:
Collapse load interaction surfaces for both the approximate and the improved solution are shown in Fig. 7. for a shell of $c = 0.4$, $\alpha = 1/20$. The agreement between the two solutions is not as good as in the previous example. This may be partly due to the fact that a shell of $c = 0.4$ is close to the border line of "shallow" and the use of the shallow shell equations may introduce some errors. Within the limitations of our theory

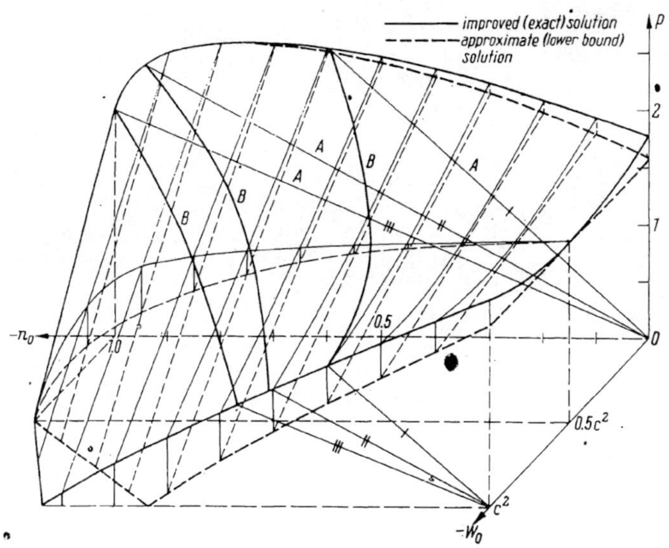

Fig. 3. Collapse load interaction surface $c = 0.1$, $\alpha = 1/20$

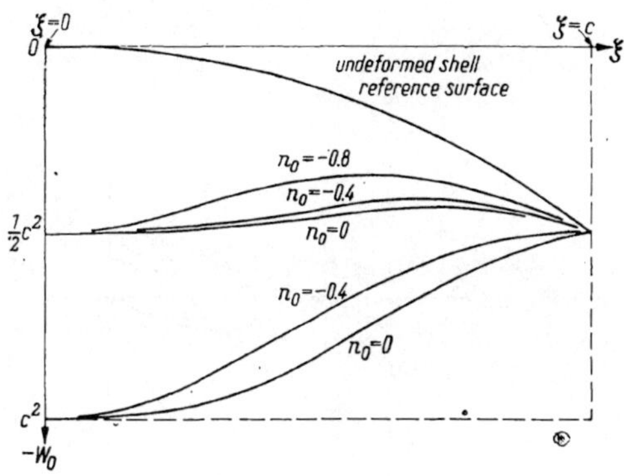

Fig. 5. Lateral displacements W, $c = 0.1$, $\alpha = 1/20$

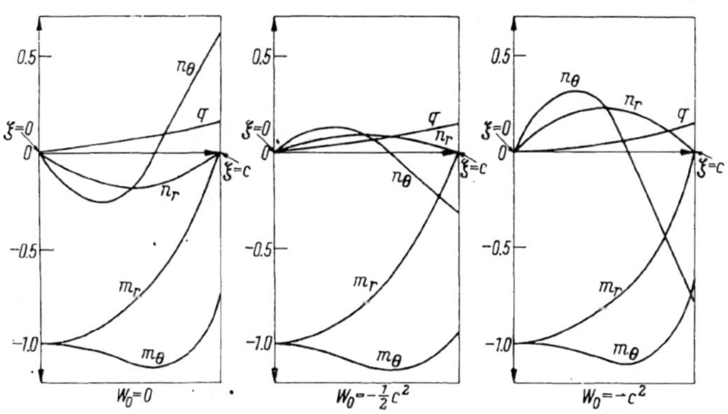

Fig. 4. Typical stress fields, $c = 0.1$, $\alpha = 1/20$, $n_0 = 0$

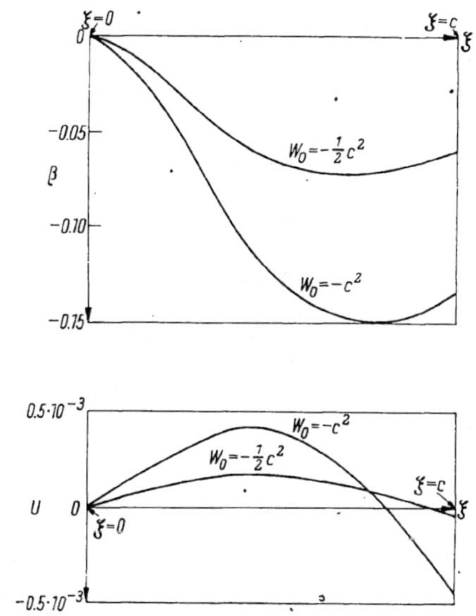

Fig. 6. Displacements β and U, $c = 0.1$, $\alpha = 1/20$, $n_0 = 0$

these shells show only "geometrical weakening" with increasing deflections. For the improved solution the n_0-P load interaction is single values for small values of W_0 and becomes double valued for increasing W_0. This is the opposite of what we observed in the shells of the previous example. Figs. 8–10 show some typical stress and displacement fields for this shell, obtained from the improved solution.

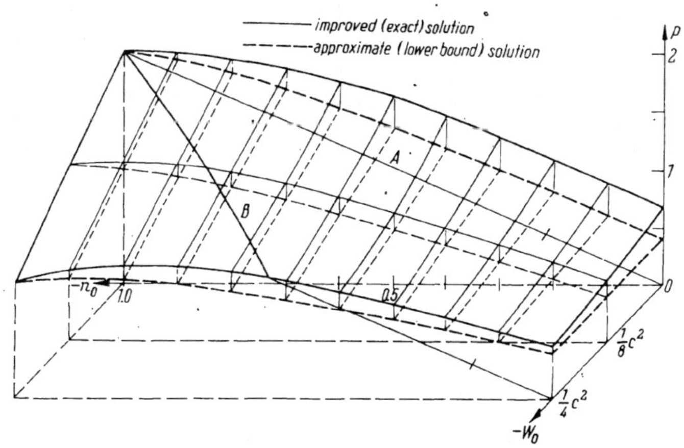

Fig. 7. Collapse load interaction surface $c = 0.4$, $\alpha = 1/20$

6. Conclusions

On the basis of our analysis we should like to suggest the following:

1. The numerical scheme used here may be useful in the analysis of problems that are extremely difficult to solve by analytical methods.
2. The approximate solution suggested gives acceptable results only for extremely shallow shells.
3. The inclusion of transverse shear terms in the analysis does not complicate the already difficult problem.
4. Since our results for the displacement field are based on the deformation theory of plasticity, experimental verification would be of interest. Nevertheless our results should be considered a first step in the solution of the title problem.

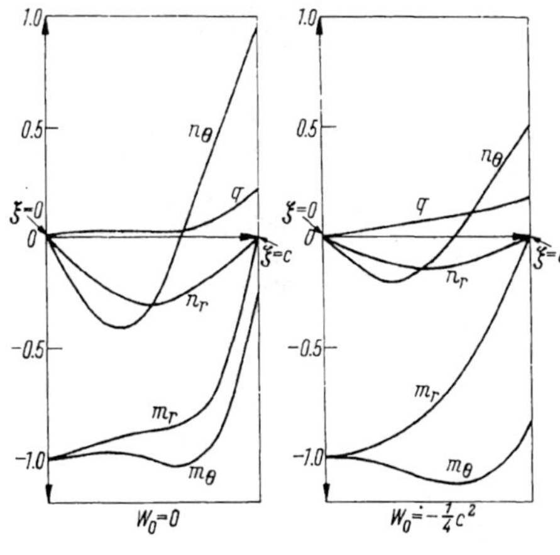

Fig. 8. Typical stress fields, $c = 0.4$, $\alpha = 1/20$, $n_0 = 0$

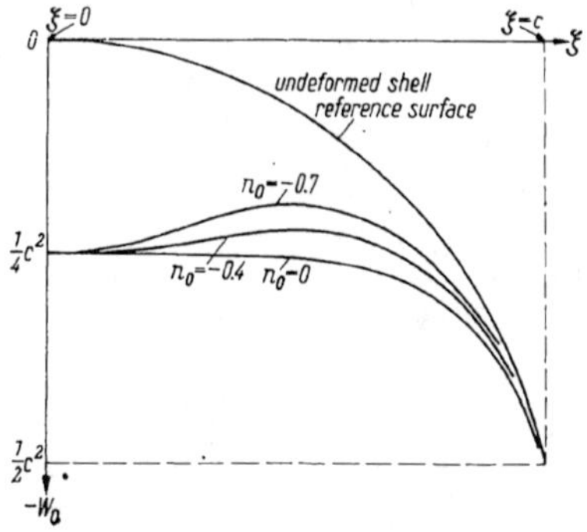

Fig. 9. Lateral displacements W, $c = 0.4$, $\alpha = 1/20$

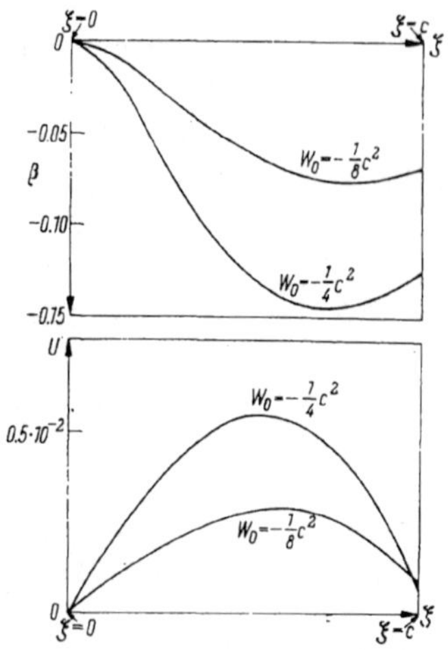

Fig. 10. Displacements β and U, $c = 0.4$, $\alpha = 1/20$, $n_0 = 0$

Acknowledgements

This work was carried out in the Department of Civil Engineering of the University of Waterloo under National Resarch Council of Canada Grant A-1582.

References

1 HODGE, P. G., Plastic Analysis of Structures, McGraw-Hill, N.Y. 1959.
2 MIKELADZE, M. SH., Estimating the carrying capacity of middly sloping shells of revolution and tension-bent round plates, Prikl. Mekh. **1** (1965).
3 SHABLII, O. N., Carrying capacity of a compression-bent mildly sloping spherical shell, Prikl. Mekh. **1** (1965).
4 KUPRIICHUK, P. F. and SHABLII, O. N., Bearing capacity of shallow variable-thickness shells of revolution. Problemy Prochnosti **3**, pp. 38–44 (1971).
5 KUPRIICHUK, P. F. and SHABLII, O. N., Determining the bearing capacity of shallow shells of revolution, Prikl. Mekh. **6** (1970).
6 SHABLII, O. N. and KUPRIICHUK, P. F., Determining the bearing capacity of shallow shells of revolution that vary in thickness, Prikl. Mekh. **5** (1969).
7 SHABLII, O. N. and KUPRIICHUK, P. F., Determining the carrying capacity of mildly sloping shells of revolution, Prikl. Mekh. **2** (1966).
8 HODGE, P. G., The collapse load of a sherical cap. Proc. 4th Midw. Conf. Solid Mech., Austin, Texas, 1959.
9 FEINBERG, S. M., Plastic flow of a shallow shell as an axisymmetric problem, PMM **4** (1957).
10 LELLEP, IA. A., Limiting equilibrium of shallow spherical shells prepared from a material with different yield stresses in tension and compression, Prikl. Mekh. **8** (1972).
11 DEKHTYAR, A. S. and VARVAK, M. SH., Carrying capacity of middly sloping shells with a central hole, Prikl. Mekh. **4** (1968).
12 DERKACH, P. KH. and SHEVCHENKO, V. P., The carrying capacity of a flat spherical shell, Inzh. Zhurnal **5** (1965).
13 DUSZEK, M., Plastic analysis of shallow spherical shells at moderately large deflections, IUTAM-Theory of Thin Shells, Edited by Niordson, Copenhagen 1967.
14 SHABLII, O. N., Large deflections of a rigid-plastic shallow spherical shell, Notes of the Summer School on "Physically and geometrically non-linear problems of the Theory of Plates and Shells", Tartu, 1966.
15 SHERBOURNE, A. N., Elastic post-buckling behaviour of a simply supported circular plate, J. Mech. Engr. Sci **3** (1961).
16 REISSNER, E., J. Math. Phys. **25**, p. 80 (1946); **25**, p. 279 (1946); **27**, p. 240 (1948); **38**, p. 16 (1959).
17 MUSHTARI, KH. M. and GALIMOV, K. Z., Non-linear Theory of Thin Elastic Shells, U.S. Dept. of Comm., N62-72586, 1957.
18 OGIBALOV, P. M., Dynamics and Strength of Shells, U.S. Dept. of Comm., TT65-50118, 1966.
19 ILYUSHIN, A. A., Plasticity, GTTI, 1948.
20 SHAPIRO, G. S., On yield surfaces for ideally plastic shells, Probl. Contin. Mech., SIAM, 1961.
21 SHROEDER, J., A general theorem for thin shells in classical plasticity, J. Math. Phys. **47**, p. 95–108 (1968).
22 ROBINSON, M., A comparison of yield surfaces for thin shells, Int. J. Mech. Sci. **13** (1971).
23 HAYDL, H. M. and SHERBOURNE, A. N., Yield surfaces for thin shells accounting for transverse shear, Arch. Mech. Stos. **25**, p. 663 (1973).
24 PRAGER, W., An Introduction to Plasticity, Addison-Wesley Co, London 1959.
25 HILL, R., The Mathematical Theory of Plasticity, Oxford Press, 1950.
26 DRUCKER, D. C., Stress-Strain relations in the plastic range – a survey of theory and experiment, O.N.R. report no. A 1151, Grad. Div. of Appl. Math., Brown Univ. 1950.
27 ILYUSHIN, A. A., Relation between the theory of Sain Venant–Levy–v. Mises and the theory of small elastic-plastic deformations, in Plastic Deformation, Edit. by H. H. Hausner, Chapter 4.
28 BUDIANSKY, B., A reassessment of deformation theories of plasticity, ASME, J. Appl. Mech., June 1959.
29 SCARBOROUGH, J. B., Numerical Mathematical Analysis, Johns Hopkins Press, 1962.
30 HAYDL, H. M. and SHERBOURNE, A. N., Rigid-Plastic collapse of compression-bent shallow shells, Int. J. Mech. Sci. **15** (1973).

Eingereicht am 3. 5. 1972, rev. Fassung 14. 11. 1972.

Anschrift: Professor Dr. A. N. SHERBOURNE and Ing. H. M. HAYDL, Department of Civil Engineering, University of Waterloo, Waterloo, Ontario, Canada

ZAMM 54, 83 – 90 (1974)

K. R. SCHNEIDER

Zur Abschätzung der Frequenz nichtlinearer Schwingungen einer Klasse gewöhnlicher Differentialgleichungen

Es wird eine obere Schranke für die Frequenz einer beliebigen periodischen Lösung der nichtlinearen Differentialgleichung $d^{(m)}x/dt^m + r_1\, d^{(m-1)}x/dt^{m-1} + \cdots + r_{m-1}\, dx/dt = f(x)$ unter der Voraussetzung abgeleitet, daß $|f'(x)| \leqq M$ für $-\infty < x < +\infty$ gilt. Die erhaltene Schranke, die unter Verwendung des Stetigkeitsmoduls der Funktion $f(x)$ abgeleitet wurde, stellt eine Funktion der Koeffizienten r_i und der Konstante M dar.

An upper bound for the frequency of any periodic solution of the nonlinear differential equation $d^{(m)}x/dt^m + r_1\, d^{(m-1)}x/dt^{m-1} + \cdots + r_{m-1}\, dx/dt = f(x)$ is derived supposing $|f'(x)| \leqq M$ for $-\infty < x < +\infty$. The bound, obtained by using the module of continuity of the function $f(x)$, is a function of the coefficients r_i and of the constant M.

Рассматривается нелинейное дифференциальное уравнение $d^{(m)}x/dt^m + r_1\, d^{(m-1)}x/dt^{m-1} + \cdots + r_{m-1}\, dx/dt = f(x)$. При условии $|f'(x)| \leqq M$, $-\infty < x < +\infty$, дается верхняя граница частоты любого периодического решения. Эта граница, полученная при помощи модуля непрерывности функции $f(x)$, зависит от коефициентов r_i и постоянного M.

1. Bei der Untersuchung zahlreicher Probleme der Regelungstechnik wird man auf eine nichtlineare Differentialgleichung der Gestalt

$$L_m[x] \equiv \frac{d^{(m)}x}{dt^m} + r_1 \frac{d^{(m-1)}x}{dt^{m-1}} + \cdots + r_{m-1} \frac{dx}{dt} = f(x)\,, \tag{1}$$

$x \in E_1$, $r_i = \text{const.}$, geführt. Die Frage nach der Existenz periodischer Lösungen von (1) und das Problem der Abschätzung ihrer Frequenz und Amplitude besitzt in diesem Zusammenhang eine große Bedeutung.

Im folgenden sollen Abschätzungen für die Frequenz ω einer beliebigen periodischen Lösung von (1) abgeleitet werden. Mit dem gleichen Problem beschäftigte sich P. SAGIROW in mehreren Arbeiten [3]–[5]. Die von ihm erhaltenen Abschätzungen wurden unter der Voraussetzung abgeleitet, daß die betrachtete periodische Lösung eine erste Harmonische mit nicht verschwindender Amplitude besitzt und die Koeffizienten des Operators $L_m[x]$ gewissen Bedingungen genügen. In den Sätzen 1 bis 3 der vorliegenden Arbeit werden obere Schranken für die Frequenz ω einer beliebigen periodischen Lösung von (1) und ohne zusätzliche Bedingungen über die Koeffizienten des Operators L_m angegeben, wobei zu bemerken ist, daß die erhaltenen Schranken besser sind als die entsprechenden Schranken bei SAGIROW [3], [4]. In den Sätzen 4 und 5 werden unter zusätzlichen Bedingungen über die Koeffizienten von L_m obere Schranken über die Frequenz einer periodischen Lösung von (1) abgeleitet. Satz 5 stimmt im wesentlichen mit einem Resultat von SAGIROW [5] überein.

2. Wir nehmen an, daß die Differentialgleichung (1) eine nichtkonstante periodische Lösung $x = p(t)$ mit der Periode T besitzt. Unter Benutzung der Kreisfrequenz $\omega = 2\,\pi/T$ und der Substitution $\tau = \omega\, t$ können wir die T-periodische Funktion $p(t)$ in eine $2\,\pi$-periodische Funktion $q(\tau) \equiv p(\tau/\omega)$ umwandeln. Über die Funktion $f(x)$ setzen wir vorerst nur voraus, daß sie über ein beliebiges Intervall quadratisch integrierbar ist; für m gelte $m \geqq 2$.

Aus unseren Voraussetzungen folgt [6], S. 76, S. 120, daß wir $q(\tau)$ als FOURIERreihe

$$q(\tau) = \frac{a_0}{2} + \sum_{n=1}^{\infty} (a_n \cos n\tau + b_n \sin n\tau) \tag{2}$$

darstellen können, die im Intervall $[0, 2\,\pi]$ gleichmäßig gegen $q(\tau)$ konvergiert. Gleichzeitig können die FOURIERreihen für die Ableitungen $\frac{d^r q}{d\tau^r}$, $1 \leqq r \leqq m$, durch gliedweise Differentiationen der Reihe (2) erhalten werden, wobei alle diese Reihen mit eventueller Ausnahme der zu $\frac{d^m q}{d\tau^m}$ gehörenden Reihe absolut und gleichmäßig gegen die entsprechenden Ableitungen konvergieren.

Aus der Eigenschaft, daß $f(x)$ quadratisch integrierbar ist, folgt, daß der Funktion $f(q(\tau))$ eindeutig eine FOURIERreihe

$$f(q(\tau)) \sim \frac{A_0}{2} + \sum_{n=1}^{\infty} (A_n \cos n\tau + B_n \sin n\tau) \tag{3}$$

zugeordnet ist. Nach Voraussetzung gilt

$$L_m[q(\tau)] = f(q(\tau))\,. \tag{4}$$

Ersetzen wir in Gleichung (4) beide Seiten durch ihre eindeutig zugeordneten FOURIERreihen, so erhalten wir durch anschließenden Koeffizientenvergleich die folgenden Beziehungen

$$0 = A_0 , \tag{5}$$

$$\left.\begin{aligned} a_n U(\omega n) + b_n V(\omega n) &= A_n , \\ - a_n V(\omega n) + b_n U(\omega n) &= B_n , \end{aligned}\right\} \quad n = 1, 2, 3, \ldots \tag{6}$$

$U(z)$ und $V(z)$ stellen dabei die Polynome

$$U(z) \equiv \sum_{k=1}^{\left[\frac{m}{2}\right]} (-1)^k r_{m-2k} z^{2k} , \qquad V(z) \equiv \sum_{k=0}^{\left[\frac{m-1}{2}\right]} (-1)^k r_{m-2k-1} z^{2k+1} \tag{7}$$

dar, wobei $[y]$ die größte in y enthaltene ganze Zahl bedeutet und $r_0 = 1$ gilt.

Die größte gemeinsame Wurzel der Polynomgleichungen $U(z) = 0$ und $V(z) = 0$ bezeichnen wir mit $z = \Omega_0$. Es gilt somit

$$U^2(\omega n) + V^2(\omega n) \neq 0 \qquad \text{für} \quad \omega > \Omega_0 , \quad n \geqq 1 . \tag{8}$$

Aus unserer Voraussetzung, daß (1) eine nichttriviale periodische Lösung $p(t)$ besitzt, folgt, daß nicht alle FOURIERkoeffzienten mit einem Index $n \geqq 1$ verschwinden. Es gibt daher ein $k \geqq 1$, so daß die Relationen

$$a_n^2 + b_n^2 = 0 \qquad \text{für} \quad 1 \leqq n \leqq k - 1 , \qquad a_k^2 + b_k^2 \neq 0 \tag{9}$$

gelten.

Aus (6) erhalten wir für $n = 1, 2, \ldots$

$$\left.\begin{aligned} a_n [U^2(\omega n) + V^2(\omega n)] &= A_n U(\omega n) - B_n V(\omega n) , \\ b_n [U^2(\omega n) + V^2(\omega n)] &= A_n V(\omega n) + B_n U(\omega n) . \end{aligned}\right\} \tag{10}$$

Daraus folgen die Beziehungen

$$(|a_n| + |b_n|) [U^2(\omega n) + V^2(\omega n)] \leqq (|A_n| + |B_n|) \big(|U(\omega n)| + |V(\omega n)|\big) \tag{11}$$

bzw.

$$(a_n^2 + b_n^2) [U^2(\omega n) + V^2(\omega n)]^2 = (A_n^2 + B_n^2) [U^2(\omega n) + V^2(\omega n)] . \tag{12}$$

Wenn wir $|A_n| + |B_n|$ bzw. $A_n^2 + B_n^2$ für $n = k, k + 1, \ldots$ durch eine Beziehung der Form

$$|A_n| + |B_n| \leqq C_1 (|a_n| + |b_n|) \tag{13}$$

bzw.

$$A_n^2 + B_n^2 \leqq C_2 (a_n^2 + b_n^2) \tag{14}$$

abschätzen könnten, so ließen sich damit aus (11) und (12) Ungleichungen über die Frequenz ω einer beliebigen periodischen Lösung von (1) ableiten. Mit diesem Problem beschäftigen wir uns im folgenden Abschnitt.

3. Wir setzen jetzt voraus, daß die Funktion $f(x)$ für $-\infty < x < +\infty$ den Bedingungen

$$f(x) \in C^{(1)} , \tag{15}$$

$$|f'(x)| \leqq M \tag{16}$$

genügt. Nach [8], S. 42ff., gelten für die FOURIERkoeffizienten A_n und B_n der Reihe (3) die Abschätzungen $(n = 1, 2, \ldots)$

$$|A_n| \leqq \frac{2}{\pi n} \bigvee_0^{2\pi} f(q(\tau)) , \qquad |B_n| \leqq \frac{2}{\pi n} \bigvee_0^{2\pi} f(q(\tau)) , \tag{17}$$

wobei $\bigvee_0^{2\pi} f(q(\tau))$ die totale Variation von $f(q(\tau))$ auf dem Intervall $[0, 2\pi]$ bedeutet, bzw.

$$|A_n| \leqq \mu\left(f, \frac{\pi}{n}\right), \qquad |B_n| \leqq \mu\left(f, \frac{\pi}{n}\right), \tag{18}$$

wobei $\mu(f, \pi/n)$ den Stetigkeitsmodul von $f(q(\tau))$ bezeichnet, der durch

$$\mu\left(f, \frac{\pi}{n}\right) = \sup_{\substack{\tau_1, \tau_2 \in [0, 2\pi] \\ |\tau_1 - \tau_2| \leqq \frac{\pi}{n}}} |f(q(\tau_1)) - f(q(\tau_2))| \tag{19}$$

definiert ist.

Unter Verwendung der Voraussetzung (16) erhalten wir nach bekannten Sätzen über die totale Variation [1], S. 241ff., bzw. direkt aus der Definitionsgleichung (19) die Abschätzungen

$$\bigvee_0^{2\pi} f(q(\tau)) \leqq M \bigvee_0^{2\pi} q(\tau) \tag{20}$$

bzw.

$$\mu\left(f(q(\tau)), \frac{\pi}{n}\right) \leqq M\, \mu\left(q(\tau), \frac{\pi}{n}\right). \tag{21}$$

Nach (2) und (9) folgt weiter

$$\bigvee_0^{2\pi} q(\tau) = \bigvee_0^{2\pi}\left(\frac{a_0}{2} + \sum_{n=k}^{\infty}(a_n \cos n\tau + b_n \sin n\tau)\right) \leqq 4\, k\, (|a_k| + |b_k|) + \bigvee_0^{2\pi} \sum_{n=k+1}^{\infty} (a_n \cos n\tau + b_n \sin n\tau)$$

$$\leqq 4\, k\, (|a_k| + |b_k|) + \sum_{n=k+1}^{\infty} \bigvee_0^{2\pi} \sqrt{a_n^2 + b_n^2}\, \sin(n\,\tau + \varphi_n) = 4\, k\, (|a_k| + |b_k|) + 4 \sum_{n=k+1}^{\infty} n\sqrt{a_n^2 + b_n^2}\,. \tag{22}$$

Für $n = 1, 2, \ldots$ und $\omega > \Omega_0$ gilt nach (12)

$$a_n^2 + b_n^2 = \frac{A_n^2 + B_n^2}{U^2(\omega\, n) + V^2(\omega\, n)}. \tag{23}$$

Unter Verwendung von (17) erhalten wir daraus

$$n\sqrt{a_n^2 + b_n^2} = n\sqrt{\frac{A_n^2 + B_n^2}{U^2(\omega\, n) + V^2(\omega\, n)}} \leqq \frac{2\sqrt{2}}{\pi}\, \frac{\bigvee_0^{2\pi} f(q(\tau))}{\sqrt{U^2(\omega\, n) + V^2(\omega\, n)}}. \tag{24}$$

Damit folgt aus (20) und (22)

$$\bigvee_0^{2\pi} f(q(\tau)) \leqq M\, 4\, k\, (|a_k| + |b_k|) + M\, \frac{8\sqrt{2}}{\pi} \bigvee_0^{2\pi} f(q(\tau)) \sum_{n=k+1}^{\infty} \frac{1}{\sqrt{U^2(\omega\, n) + V^2(\omega\, n)}}. \tag{25}$$

Führen wir die Funktion

$$\alpha_{k+1}(\omega) \equiv \frac{8\sqrt{2}\, M}{\pi} \sum_{n=k+1}^{\infty} \frac{1}{\sqrt{U^2(\omega\, n) + V^2(\omega\, n)}} \tag{26}$$

ein, so sieht man unter Verwendung der Definitionen (7) für $U(\omega\, n)$ und $V(\omega\, n)$, daß $\alpha_{k+1}(\omega)$ für $m \geqq 2$ und $\omega \to \infty$ der Relation

$$\alpha_{k+1}(\omega) = o\left(\frac{1}{\omega}\right) \tag{27}$$

genügt. Es existiert somit eine größte positive Wurzel der Gleichung $\alpha_{k+1}(\omega) - 1 = 0$, die wir mit Ω_{k+1} bezeichnen. Trivialerweise gilt $\Omega_{k+1} > \Omega_0$. Somit haben wir für $\omega > \Omega_{k+1}$

$$\alpha_{k+1}(\omega) < 1\,. \tag{28}$$

Unter dieser Voraussetzung erhalten wir aus (25) die Abschätzung

$$\bigvee_0^{2\pi} f(q(\tau)) \leqq \frac{4\, k\, M\, (|a_k| + |b_k|)}{1 - \alpha_{k+1}(\omega)}. \tag{29}$$

Analog zu dieser Herleitung haben wir nach (9), (18), (19) und (23)

$$\mu\left(q, \frac{\pi}{k}\right) = \mu\left[\frac{a_0}{2} + a_k \cos k\tau + b_k \sin k\tau + \sum_{n=k+1}^{\infty}(a_n \cos n\tau + b_n \sin n\tau), \frac{\pi}{k}\right] \leqq 2\,(|a_k| + |b_k|) +$$

$$+ \mu\left(\sum_{n=k+1}^{\infty} \sqrt{a_n^2 + b_n^2}\, \sin(n\tau + \varphi_n), \frac{\pi}{k}\right) \leqq 2\,(|a_k| + |b_k|) + 2 \sum_{n=k+1}^{\infty} \sqrt{\frac{A_n^2 + B_n^2}{U^2(\omega\, n) + V^2(\omega\, n)}}$$

$$\leqq 2\,(|a_k| + |b_k|) + 2\sqrt{2}\, M\, \mu\left(q, \frac{\pi}{k}\right) \sum_{n=k+1}^{\infty} \frac{1}{\sqrt{U^2(\omega\, n) + V^2(\omega\, n)}}. \tag{30}$$

Setzen wir

$$\beta_{k+1}(\omega) \equiv 2\sqrt{2}\, M \sum_{n=k+1}^{\infty} \frac{1}{\sqrt{U^2(\omega\, n) + V^2(\omega\, n)}} \tag{31}$$

und beachten wir die Beziehung

$$\beta_{k+1}(\omega) = \frac{\pi}{4}\, \alpha_{k+1}(\omega)\,, \tag{32}$$

so gilt für $\omega > \Omega_{k+1}$ die Ungleichung $\beta_{k+1}(\omega) < 1$. Für $\omega > \Omega_{k+1}$ erhalten wir folglich aus (30) und (31)

$$\mu\left(q(\tau), \frac{\pi}{k}\right) \leqq \frac{2\,(|a_k| + |b_k|)}{1 - \beta_{k+1}(\omega)}. \tag{33}$$

Unter den Voraussetzungen (15) und (16) erhalten wir für $\omega > \Omega_{k+1}$ aus (17) und (29) für $n = k$ die Abschätzungen

$$|A_k| + |B_k| \leqq \frac{16\,M}{\pi\,(1 - \alpha_{k+1}(\omega))}\,(|a_k| + |b_k|)\,, \tag{34}$$

$$A_k^2 + B_k^2 \leqq 8\left(\frac{4\,M}{\pi\,(1 - \alpha_{k+1}(\omega))}\right)^2 (|a_k| + |b_k|)^2 \leqq \left(\frac{16\,M}{\pi\,(1 - \alpha_{k+1}(\omega))}\right)^2 (a_k^2 + b_k^2)\,. \tag{35}$$

Analog erhalten wir für $\omega > \Omega_{k+1}$ aus (18), (21) und (33)

$$|A_k| + |B_k| \leqq \frac{4\,M}{1 - \beta_{k+1}(\omega)}\,(|a_k| + |b_k|)\,, \quad A_k^2 + B_k^2 \leqq \left(\frac{4\,M}{1 - \beta_{k+1}(\omega)}\right)^2 (a_k^2 + b_k^2)\,. \tag{36), (37}$$

Die erhaltenen Abschätzungen sind alle vom Typ (13) bzw. (14). Wie man mittels der Relation (32) leicht nachprüfen kann, sind die Abschätzungen (36), (37), die unter Verwendung des Stetigkeitsmoduls erhalten wurden, besser als die Abschätzungen (34), (35), die unter Verwendung der totalen Variation abgeleitet wurden. Durch Einsetzen von (36) und (37) in (11) bzw. (12) erhalten wir nach (9) folgende Ungleichungen

$$U^2(k\,\omega) + V^2(k\,\omega) \leqq \frac{4\,M}{1 - \beta_{k+1}(\omega)}\,(|U(k\,\omega)| + |V(k\,\omega)|)\,; \tag{38}$$

$$U^2(k\,\omega) + V^2(k\,\omega) \leqq \left(\frac{4\,M}{1 - \beta_{k+1}(\omega)}\right)^2. \tag{39}$$

Diese Abschätzungen haben dabei folgende Bedeutung. Besitzt das System (1) unter den Voraussetzungen (15) und (16) eine periodische Lösung mit der Frequenz $\omega = \omega^* > \Omega_{k+1}$, so genügt ω^* den Ungleichungen (38) und (39). Wir bemerken noch, daß sich die Ungleichung (38) aus der Ungleichung (39) ableiten läßt. Aus (39) folgt zunächst

$$\sqrt{U^2(k\,\omega) + V^2(k\,\omega)} \leqq \frac{4\,M}{1 - \beta_{k+1}(\omega)}\,. \tag{40}$$

Wegen

$$\sqrt{U^2(k\,\omega) + V^2(k\,\omega)} \geqq \frac{U^2(k\,\omega) + V^2(k\,\omega)}{|U(k\,\omega)| + |V(k\,\omega)|} \tag{41}$$

folgt aus (40) die Ungleichung (38). Unsere bisherigen Untersuchungen können wir somit in folgendem Satz zusammenfassen:

Satz 1. *Bezeichnen wir mit ω_k die größte positive Wurzel der Gleichung*

$$U^2(\omega\,k) + V^2(\omega\,k) = \left(\frac{4\,M}{1 - \beta_{k+1}(\omega)}\right)^2, \tag{42}$$

so stellt unter den Voraussetzungen (15) *und* (16) *die Zahl ω_k eine obere Schranke für die Frequenz ω^* einer beliebigen nichttrivialen periodischen Lösung $p(t)$ von* (1) *dar.*

Bei einer praktischen Anwendung von Satz 1 stört zunächst noch die unbekannte Zahl k. Zur Umgehung dieser Schwierigkeit ersetzen wir die Ungleichung (39) durch die schwächere Ungleichung

$$U^2(k\,\omega) + V^2(k\,\omega) \leqq \left(\frac{4\,M}{1 - \beta_2(\omega)}\right)^2. \tag{43}$$

Aus Satz 1 folgt somit

Korollar 1.1: *Bezeichnen wir mit $\overline{\omega}$ die größte positive Wurzel der Gleichungen*

$$U^2(\omega\,k) + V^2(\omega\,k) = \left(\frac{4\,M}{1 - \beta_2(\omega)}\right)^2, \qquad k = 1, 2, \ldots, \tag{44}$$

so stellt unter den Voraussetzungen (15) *und* (16) *die Zahl $\overline{\omega}$ eine obere Schranke für die Frequenz ω^* einer beliebigen periodischen Lösung $p(t)$ von* (1) *dar.*

Im folgenden wenden wir dieses Korollar auf die Differentialgleichung

$$\frac{d^2x}{dt^2} = f(x) \tag{45}$$

an, wobei $f(x)$ den Voraussetzungen (15) und (16) genügen soll. Nach (7) haben wir dann $U(z) = -z^2$, $V(z) \equiv 0$, $\Omega_0 = 0$. Aus (31) folgt somit

$$\beta_2(\omega) = 2\sqrt{2}\,M \sum_{n=2}^{\infty} \frac{1}{(\omega\,n)^2} = \frac{2\sqrt{2}\,M}{\omega^2}\left(\frac{\pi^2}{6} - 1\right). \tag{46}$$

In diesem Fall haben die Gleichungen (44) die Gestalt

$$\omega^2 k^2\left[1-\frac{2\sqrt{2}\,M}{\omega^2}\left(\frac{\pi^2}{6}-1\right)\right]=4\,M\,,\qquad k=1,2,\ldots,$$

also

$$\omega^2=2\,M\left[\sqrt{2}\left(\frac{\pi^2}{6}-1\right)+\frac{2}{k^2}\right].$$

Folglich gilt

$$\bar{\omega}=\omega_1<2{,}42\sqrt{M}\,.$$

Nach Korollar 1.1 haben wir somit das Resultat:

Satz 2. *Unter den Voraussetzungen (15) und (16) gilt für die Periode T einer beliebigen periodischen Lösung der Gleichung (45) die Relation*

$$T\geqq\frac{2\,\pi}{2{,}42\sqrt{M}}\,. \tag{47}$$

Dieses Ergebnis stimmt mit einem allgemeinen qualitativen Resultat über die Periode freier ungedämpfter Schwingungen überein [2], S. 107.

4. Im folgenden Abschnitt zeigen wir, wie man die durch Satz 1 bestimmte obere Schranke für die Frequenz einer beliebigen periodischen Lösung von (1) durch eine gewisse Phasennormierung weiter verbessern kann, ohne zusätzliche Voraussetzungen zu treffen. Aus der Eigenschaft, daß die rechte Seite von (1) nicht explizit von t abhängt, folgt, daß gleichzeitig mit $p(t)$ auch $p(t+\varphi)$ eine Lösung von (1) ist, wobei φ eine beliebige reelle Zahl (Phasenverschiebung) bedeutet.

Es läßt sich leicht nachprüfen, daß man durch geeignete Wahl der Konstanten φ erreichen kann, daß die zu $p(t+\varphi)$ gehörende 2π-periodische Funktion $q(\tau+\omega\varphi)$ eine FOURIERentwicklung der Gestalt

$$\tilde{q}(\tau)=\frac{a_0}{2}+\tilde{a}_k\cos k\tau+\sum_{n=k+1}^{\infty}(\tilde{a}_n\cos n\tau+\tilde{b}_n\sin n\tau) \tag{48}$$

bzw. der Gestalt

$$\bar{q}(\tau)=\frac{a_0}{2}+\bar{b}_k\sin k\tau+\sum_{n=k+1}^{\infty}(\bar{a}_n\cos n\tau+\bar{b}_n\sin n\tau) \tag{49}$$

besitzt.

Stellen wir $f(\tilde{q}(\tau))$ als FOURIERreihe

$$f(\tilde{q}(\tau))\sim\frac{\tilde{A}_0}{2}+\sum_{n=1}^{\infty}(\tilde{A}_n\cos n\tau+\tilde{B}_n\sin n\tau) \tag{50}$$

dar und ersetzen wir wieder beide Seiten von (1) durch die entsprechenden FOURIERreihen, so erhalten wir durch Koeffizientenvergleich in Analogie zu (5) und (6) das System

$$\begin{aligned}&0=\tilde{A}_0\,,\\&\tilde{a}_k\,U(\omega\,k)=\tilde{A}_k\,,\qquad -\tilde{a}_k\,V(\omega\,k)=\tilde{B}_k\,,\\&\tilde{a}_n\,U(\omega\,n)+\tilde{b}_n\,V(\omega\,n)=\tilde{A}_n\,,\qquad -\tilde{a}_n\,V(\omega\,n)+\tilde{b}_n\,U(\omega\,n)=\tilde{B}_n\,,\qquad n=k+1,k+2,\ldots\end{aligned} \tag{51}$$

Unter Verwendung der aus (33) folgenden Relation

$$\mu\left(\tilde{q}(\tau),\frac{\pi}{k}\right)\leqq\frac{2\,|\tilde{a}_k|}{1-\beta_{k+1}(\omega)}$$

erhalten wir nach (18) und (21) für $n=k$

$$|\tilde{A}_k|\leqq\frac{2\,M\,|\tilde{a}_k|}{1-\beta_{k+1}(\omega)}\,,\qquad|\tilde{B}_k|\leqq\frac{2\,M\,|\tilde{a}_k|}{1-\beta_{k+1}(\omega)}\,. \tag{52}$$

Mit diesen Abschätzungen über die Koeffizienten $\tilde{A}_k$ und $\tilde{B}_k$ läßt sich folgender Satz formulieren, der eine Verbesserung von Satz 1 darstellt.

Satz 3: *Bezeichnen wir mit ω_U bzw. ω_V die größte positive Nullstelle der Gleichungen*

$$|U(\omega\,k)|=\frac{2\,M}{1-\beta_2(\omega)}\,,\qquad|V(\omega\,k)|=\frac{2\,M}{1-\beta_2(\omega)}\,,\qquad k=1,2,\ldots, \tag{53}$$

wobei $\omega_U = +\infty$ ($\omega_V = +\infty$) gesetzt werden soll, falls $U(z)$ ($V(z)$) identisch verschwindet, so stellt unter den Voraussetzungen (15) *und* (16) *die Zahl*

$$\tilde{\omega} = \operatorname{Min}(\omega_U, \omega_V)$$

eine obere Schranke für die Frequenz ω einer beliebigen nichttrivialen periodischen Lösung von (1) *dar.*

Bemerkung 3.1: Man gelangt zum selben Resultat, wenn anstelle der Entwicklung (48) die FOURIERentwicklung (49) benutzt wird.

Bemerkung 3.2: Satz 1 kann als Folgerung aus Satz 3 erhalten werden.

Die Anwendung dieses Satzes auf die Differentialgleichung (45) liefert unter Verwendung von (46) das Resultat

Satz 2': *Unter den Voraussetzungen* (15) *und* (16) *gilt für die Periode T einer beliebigen periodischen Lösung der Gleichung* (45) *die Relation*

$$T \geqq \frac{2\pi}{\tilde{\omega}} \geqq \frac{2\pi}{1{,}96\sqrt{M}}.$$

Diese Abschätzung ist offensichtlich besser als die von (47).

Eine weitere Möglichkeit, eine Abschätzung für die Koeffizienten $\tilde{B}_k$ zu erhalten, besteht in folgendem: Für den Koeffizienten $\tilde{B}_k$ gilt

$$\tilde{B}_k = \frac{1}{\pi}\int_0^{2\pi} f(\tilde{q}(\tau)) \sin k\tau \, d\tau . \tag{54}$$

Nach (48) und (15) erhalten wir

$$\begin{aligned}\int_0^{2\pi} f(\tilde{q}(\tau)) \sin k\tau\, d\tau &= \int_0^{2\pi} f\left[\frac{a_0}{2} + \tilde{a}_k \cos k\tau + \sum_{n=k+1}^{\infty} (\tilde{a}_n \cos n\tau + \tilde{b}_n \sin n\tau)\right] \sin k\tau\, d\tau \\ &= \int_0^{2\pi} f\left(\frac{a_0}{2} + \tilde{a}_k \cos k\tau\right) \sin k\tau\, d\tau + \int_0^{2\pi} (\Sigma_{k+1}) f'\left(\frac{a_0}{2} + \tilde{a}_k \cos k\tau + \vartheta\, \Sigma_{k+1}\right) \sin k\tau\, d\tau ,\end{aligned} \tag{55}$$

wobei Σ_{k+1} zur Abkürzung für $\sum_{n=k+1}^{\infty} (\tilde{a}_n \cos n\tau + \tilde{b}_n \sin n\tau)$ gesetzt wurde und ϑ eine Zahl zwischen 0 und 1 bedeutet. Wegen

$$\int_0^{2\pi} f\left(\frac{a_0}{2} + \tilde{a}_k \cos k\tau\right) \sin k\tau\, d\tau = 0$$

und (16) folgt aus (54) und (55) weiter

$$|\tilde{B}_k| \leqq M \max_{0 \leqq \tau \leqq 2\pi} |\Sigma_{k+1}| \frac{4}{\pi}. \tag{56}$$

Zur Abschätzung des zweiten Faktors verwenden wir die Relationen (18), (23), (31) und (39), wobei wir beachten, daß $\tilde{b}_k = 0$ gilt. Man erhält

$$\begin{aligned}\max_{0 \leqq \tau \leqq 2\pi} |\Sigma_{k+1}| &\leqq \sum_{n=k+1}^{\infty} \sqrt{\tilde{a}_n^2 + \tilde{b}_n^2} = \sum_{n=k+1}^{\infty} \sqrt{\frac{\tilde{A}_n^2 + \tilde{B}_n^2}{U^2(\omega n) + V^2(\omega n)}} \\ &\leqq \sqrt{2}\, \mu\left(f(\tilde{q}(\tau)), \frac{\pi}{k}\right) \sum_{n=k+1}^{\infty} \frac{1}{\sqrt{U^2(\omega n) + V^2(\omega n)}} \leqq \frac{|\tilde{a}_k|}{M} \frac{\beta_{k+1}(\omega)}{1 - \beta_{k+1}(\omega)}.\end{aligned}$$

Damit folgt aus (56)

$$|\tilde{B}_k| \leqq |\tilde{a}_k| \frac{2}{\pi} \frac{\beta_{k+1}(\omega)}{1 - \beta_{k+1}(\omega)} \leqq |\tilde{a}_k| \frac{2}{\pi} \frac{\beta_2(\omega)}{1 - \beta_2(\omega)}.$$

Unter Verwendung der dritten Gleichung von (51) erhalten wir damit folgendes Resultat: Falls die Differentialgleichung (1) unter den Voraussetzungen (15) und (16) eine nichttriviale periodische Lösung besitzt, so genügt ihre Frequenz der Ungleichung

$$|V(\omega k)| \leqq \frac{2}{\pi} \frac{\beta_2(\omega)}{1 - \beta_2(\omega)}.$$

Im Falle $V(z) \not\equiv 0$ kann daraus eine obere Schranke über die zugehörige Schwingungsdauer ermittelt werden.

Satz 4: *Unter der Voraussetzung* $V(z) \not\equiv 0$ *und den Bedingungen* (15) *und* (16) *stellt die größte positive Wurzel der Gleichungen*

$$|V(\omega k)| = \frac{2}{\pi} \frac{\beta_2(\omega)}{1 - \beta_2(\omega)}, \qquad k = 1, 2, \ldots, \tag{57}$$

eine obere Schranke für die Frequenz einer beliebigen nichttrivialen periodischen Lösung von (1) *dar.*

Bemerkung 4.1: Man gelangt zu demselben Resultat, wenn anstelle von (48) die Entwicklung (49) benutzt wird.

5. Im folgenden werden wir unter Voraussetzungen über $f(x)$, die schwächer sind als in den Abschnitten 3 und 4, eine weitere obere Schranke über die Frequenz ω^* einer beliebigen periodischen Lösung von (1) ableiten, die in gewissen Fällen die bestmögliche Schranke ist.

Wir setzen voraus, daß $f(x)$ über ein beliebiges Intervall quadratisch integrierbar ist und daß (1) eine T-periodische nichtkonstante Lösung $x = p(t)$ besitzt, der wiederum die 2π-periodische Funktion $q(\tau)$ entsprechen soll. m sei größer oder gleich 2. Bezeichnen wir mit $F(x)$ eine Stammfunktion von $f(x)$, so gilt

$$\int_0^{2\pi} \frac{dF(q(\tau))}{d\tau} d\tau = F(q(\tau)) \Big|_0^{2\pi} = 0 = \int_0^{2\pi} f(q(\tau)) \frac{dq}{d\tau} d\tau . \tag{58}$$

Unter unseren Voraussetzungen sind $f(q(\tau))$ und $dq(\tau)/d\tau$ bezüglich τ 2π-periodisch und quadratisch integrierbar. Die zu $f(q(\tau))$ gehörende FOURIERreihe sei

$$f(q(\tau)) \sim \frac{1}{2} A_0 + \sum_{n=1}^{\infty} (A_n \cos n\tau + B_n \sin n\tau) . \tag{59}$$

Nach einem allgemeinen Satz [7], S. 82, gilt dann für $0 \leqq x \leqq 2\pi$

$$\int_0^x f(q(\tau)) \frac{dq(\tau)}{d\tau} d\tau = \frac{1}{2} A_0 \int_0^x \frac{dq(\tau)}{d\tau} d\tau + \sum_{n=1}^{\infty} \left\{ A_n \int_0^x \frac{dq(\tau)}{d\tau} \cos n\tau \, d\tau + B_n \int_0^x \frac{dq(\tau)}{d\tau} \sin n\tau \, d\tau \right\}, \tag{60}$$

wobei die rechtsstehende Reihe für die $0 \leqq x \leqq 2\pi$ absolut und gleichmäßig konvergiert. Unter unseren Voraussetzungen ($m \geqq 2$) folgt aus (2)

$$\frac{dq(\tau)}{d\tau} = \sum_{n=1}^{\infty} n(-a_n \sin n\tau + b_n \cos n\tau) , \tag{61}$$

wobei diese Reihe für $0 \leqq \tau \leqq 2\pi$ absolut und gleichmäßig konvergiert. Somit ergeben sich aus (61) die Beziehungen

$$\begin{gathered} \int_0^{2\pi} \frac{dq(\tau)}{d\tau} d\tau = 0 , \\ \int_0^{2\pi} \frac{dq(\tau)}{d\tau} \cos n\tau \, d\tau = \pi n b_n , \qquad \int_0^{2\pi} \frac{dq(\tau)}{d\tau} \sin n\tau \, d\tau = -\pi n a_n . \end{gathered} \tag{62}$$

Setzen wir in (60) $x = 2\pi$ und beachten die Relationen (62) und (58), so erhalten wir

$$\sum_{n=1}^{\infty} n (A_n b_n - B_n a_n) = 0 , \tag{63}$$

wobei diese Reihe absolut konvergiert. Ersetzen wir in (63) A_n und B_n durch die Beziehungen (6), so folgt

$$\sum_{n=1}^{\infty} n (a_n^2 + b_n^2) V(n\omega) = 0 , \tag{64}$$

wobei $V(z)$ in (7) definiert ist. Wir haben somit den

Satz 5: *Unter der Voraussetzung, daß* $f(x)$ *quadratisch integrierbar ist und* $m \geqq 2$ *gilt, genügt die Frequenz* ω^* *einer beliebigen periodischen Lösung von* (1) *der Gleichung* (64).

Aus diesem Satz erhalten wir unmittelbar folgende Abschätzung über ω^*:

Satz 6: *Unter den Voraussetzungen*

1. $f(x)$ *ist über ein beliebiges Intervall quadratisch integrierbar;*
2. $m \geqq 2$;

3. $V(z) \not\equiv 0$; *die Gleichung*

$$V(z) = 0 \tag{65}$$

besitzt eine positive Nullstelle

stellt die größte positive Nullstelle Ω_m von (65) eine obere Schranke für die Frequenz ω^* einer beliebigen nichtkonstanten periodischen Lösung von (1) dar.

Beweis: Wir nehmen an, (1) besitzt eine periodische Lösung mit einer Frequenz $\omega^* > \Omega_m$. Da $V(z)$ nicht identisch verschwindet und Ω_m die größte positive Nullstelle von (65) darstellt, gilt $|V(\omega^*)| > V(\Omega_m) = 0$ für alle n, wobei $V(n\,\omega^*)$ für alle n dasselbe Vorzeichen besitzt. Aus der Eigenschaft, daß für eine nichtkonstante periodische Lösung von (1) eine Zahl $k \geqq 1$ existiert, so daß $a_k^2 + b_k^2 \neq 0$ ist, folgt somit

$$\left|\sum_{n=1}^{\infty} n\,(a_n^2 + b_n^2)\,V(\omega^*\,n)\right| = \sum_{n=1}^{\infty} n\,(a_n^2 + b_n^2)\,|V(n\,\omega^*)| \geqq k\,(a_k^2 + b_k^2)\,|V(k\,\omega^*| \neq 0\,,$$

also ein Widerspruch zur Relation (64).

Bemerkung 6.1: Die in Satz 6 angegebene obere Schranke kann durch keinen konstanten Faktor verschärft werden. Betrachten wir z. B. die lineare Differentialgleichung

$$\dddot{x} + \ddot{x} + \dot{x} = -x\,, \tag{66}$$

so erhalten wir nach (7) $V(z) \equiv z - z^3$. Nach Satz 6 erhalten wir für die Frequenz ω einer beliebigen nichttrivialen periodischen Lösung von (66) die Abschätzung $\omega \leqq 1$. Wie wir aber unmittelbar nachprüfen können, stellt die Funktion $x(t) = a \cos t + b \sin t$ eine periodische Lösung von (66) mit der Frequenz $\omega = 1$ dar.

Bemerkung 6.2: Bemerkenswert an Satz 6 ist, daß unter seinen Voraussetzungen eine obere Schranke für die Frequenz einer beliebigen nichtkonstanten periodischen Lösung angegeben werden kann, die nur von den Koeffizienten des Operators L_m, nicht aber von $f(x)$ abhängt.

Bemerkung 6.3: Ein zum Satz 6 analoges Resultat, bei dem nur die Koeffizienten von $U(z)$ eine Rolle spielen, kann es nicht geben, wie Satz 2 im Falle $m = 2$ zeigt.

Bemerkung 6.4: Satz 6 stimmt im wesentlichen mit einem von SAGIROW [5] angegebenen Resultat überein.

Abschließend soll noch darauf hingewiesen werden, daß die abgeleiteten Resultate auch auf Gleichungen vom Typ $L_m[x] = f(x, \dot{x})$ oder noch allgemeinere Gleichungen übertragen werden können.

Literatur

1 NATANSON, I. P., Theorie der Funktionen einer reellen Veränderlichen, Berlin 1954.
2 REISSIG, R., SANSONE, G., and CONTI, R., Qualitative Theorie nichtlinearer Differentialgleichungen, Rom 1963.
3 SAGIROW, P., Zur Frage der Fehlerabschätzung beim Verfahren der harmonischen Balance, ZAMM **40**, S. 456–463 (1960).
4 SAGIROW, P., Über periodische Lösungen nichtlinearer Differentialgleichungen und über eine Beziehung zwischen Amplitude und Frequenz, ZAMM **41**, S. 110–114 (1961).
5 SAGIROW, P., Eine obere Schranke für die Frequenzen periodischer Lösungen von nichtlinearen Differentialgleichungen, ZAMM **41**, s. T. 59, 60 (1961).
6 TOLSTOW, G. P., Fourierreihen, Berlin 1955.
7 TRICOMI, F. G., Vorlesungen über Orthogonalreihen, Berlin 1955.
8 ZYGMUND, A., Trigonometrical Series, Vol. I, Cambridge 1959.

Eingereicht am 13. 3. 1973

Anschrift: Dr. KLAUS R. SCHNEIDER, Zentralinstitut für Mathematik und Mechanik der Akademie der Wissenschaften der DDR, 108 Berlin, Mohrenstr. 39, DDR

ZAMM 54, 91 – 101 (1974)

R. M. Markytan

Anwendung einer indirekten Methode zur Berechnung ebener Unterschallströmungen auf die Berechnung der ebenen, stationären, inkompressiblen Strömung durch Gitter starker Umlenkung*)

Durch Beltrami-Transformation der Koordinaten auf neue unabhängige Veränderliche geht das System der gasdynamischen Gleichung und der Gleichung der Drehungsfreiheit in ein Cauchy-Riemann-System über. Von diesem wird eine von den neuen Veränderlichen abhängige numerische Lösung für die Geschwindigkeiten und Strömungswinkel angegeben. Die Koordinaten der Strömungsebene ergeben sich aus einer Integration des Beltrami-Systems. Ein dem gesamten Verfahren auferlegter Iterationsprozeß führt die bei dieser Darstellungsweise nicht selbstverständliche Schließung der Gitterprofile herbei.

The Beltrami transformation is used to convert the physical coordinates into new independent variables. The system of the gasdynamic equation and the equation of irrotational flow is thereby converted into a Cauchy-Riemann system. A numerical solution for the velocities and flow angles, depending on the new variables, is derived. The physical coordinates of the flow plane are obtained by an integration of the Beltrami system. This method usually does not give a connected profile at the trailing edge. Thus a process of iteration is imposed on the complete procedure to ensure the closure of the cascade profiles.

Преобразованием координат по Бельтрами на новые независимые переменные система газодинамического уравнения и уравнения об отсутствии вращения переходит в систему Коши-Риманна. Для скоростей и углов потока этой системы приводится числовое решение в зависимости от новых переменных. Координаты плоского течения получены интегрированием системы Бельтрами. Итерационный процесс, наложенный на весь метод, приводит к неочевидному для этого способа подхода замыканию профилей решётки.

0. Einleitung

Die bisher aufgestellten Gittertheorien lassen sich bezüglich ihrer Betrachtungsweise in drei Gruppen einteilen. Es sind dies die „Singularitätenverfahren", die „Kanaltheorien" und die Verfahren der „konformen Abbildung". Abgesehen von dieser Einteilung unterscheidet man in der Gittertheorie, wie bekannt ist, zwischen der „I. und der II. Hauptaufgabe".

Die vorliegende Arbeit, die einen Beitrag zur theoretischen Ermittlung der ebenen, stationären, inkompressiblen Strömung durch Gitter starker Umlenkung leisten soll, ist der II. Hauptaufgabe — allerdings nicht ohne Vorbehalt — zuzuordnen (eine unter demselben Vorbehalt zu verstehende Erfüllung der I. Hauptaufgabe würde nur unwesentliche Änderungen erfordern, die an den betreffenden Stellen sofort klar werden). Der von zwei benachbarten Schaufelprofilen und ihren Verlängerungen durch die vorderen und hinteren Staupunktsstromlinien gebildete ebene Strömungskanal wird zunächst auf einen beiderseits unendlich langen, zur Abszisse einer Vergleichsebene parallelen Streifen abgebildet. Die in dieser Ebene erhaltene Lösung wird schließlich in die Strömungsebene rücktransformiert, wobei jedoch erst ein dem gesamten Verfahren auferlegter Iterationsprozeß eine geschlossene Profilkontur sichert, bei dem sich die vorgegebene Profilwinkelverteilung, d. h. die Profilkontur, (bzw. die vorgegebene Geschwindigkeitsverteilung auf dem Profil) geringfügig ändert.

Während einerseits die Forderung nach starker Umlenkung der Strömung und die daraus folgende enge Gitterteilung auf eine Kanaltheorie schließen lassen, sehen andererseits jedoch die erhaltenen Gleichungen über das mathematisch singuläre Verhalten hinaus denen der Singularitätenverfahren sehr ähnlich. Anstelle von Wirbel-, Quell- und Senkbelegungen müssen wir allerdings hier von „Winkelbelegungen" bzw. „Belegungen mit einer Geschwindigkeitsfunktion" sprechen, welche am Beginn der Arbeit definiert wird. Außerdem sei erwähnt, daß die oben beschriebene Abbildung konform ist.

Die physikalischen Grundlagen der benutzten Theorie wurden von Gretler [1], [2] übernommen, dessen Arbeiten wiederum auf einem Aufsatz von Christianowitsch [3] aufbauen.

1. Grundlegende Gleichungen

Eine ebene Potentialströmung wird durch die gasdynamische Gleichung und durch die Gleichung der Drehungsfreiheit beschrieben:

$$(c^2 - u^2)\frac{\partial u}{\partial x} - u\,v\left(\frac{\partial u}{\partial y} + \frac{\partial v}{\partial x}\right) + (c^2 - v^2)\frac{\partial v}{\partial y} = 0\,, \qquad \frac{\partial u}{\partial y} - \frac{\partial v}{\partial x} = 0\,. \tag{1.1}$$

c ist die örtliche Schallgeschwindigkeit des Strömungsmediums.

Führt man statt der Geschwindigkeitskomponenten $u(x, y)$ und $v(x, y)$ den Betrag des Geschwindigkeitsvektors $\boldsymbol{W}$, nämlich $W(x, y)$, den Neigungswinkel $\vartheta(x, y)$ zwischen $\boldsymbol{W}$ und der positiven x-Achse sowie die

*) Diese Arbeit entstand auf Anregung von Herrn Professor Dr. K. Oswatitsch, Vorstand des Institutes für Strömungslehre der Technischen Hochschule Wien.

Potentialfunktion $\Phi(x, y)$ und die Stromfunktion $\Psi(x, y)$ ein, so erhält man auf Grund von

$$u = \frac{\partial \Phi}{\partial x} = \frac{1}{\varrho} \frac{\partial \Psi}{\partial y}, \qquad v = \frac{\partial \Phi}{\partial y} = -\frac{1}{\varrho} \frac{\partial \Psi}{\partial x}$$

und $M = W/c$ mit M als der örtlichen MACHzahl und ϱ als der örtlichen Dichte des Strömungsmediums aus dem System (1.1) das System

$$(1 - M^2) \frac{1}{\varrho W} \frac{\partial W}{\partial \Phi} + \frac{\partial \vartheta}{\partial \Psi} = 0, \qquad \frac{\varrho}{W} \frac{\partial W}{\partial \Psi} - \frac{\partial \vartheta}{\partial \Phi} = 0. \tag{1.2}$$

W und ϑ wurden dabei als Funktionen der „natürlichen" Koordinaten Φ und Ψ aufgefaßt.

Das Gleichungssystem (1.2) wird nun durch Einführung der Geschwindigkeitsfunktion q, der sogenannten „TSCHAPLYGIN-Variablen",

$$q = \int_{W_b}^{W} \sqrt{1 - M^2} \frac{dW}{W} \tag{1.3}$$

und der Funktion $K(q)$, einer dimensionslosen Größe,

$$K = \frac{\varrho_b}{\varrho} \sqrt{1 - M^2} \tag{1.4}$$

symmetrisch gemacht, wobei sich W_b und ϱ_b auf einen beliebigen Bezugszustand beziehen:

$$K(q) \frac{\partial q}{\partial \Phi} + \frac{\partial \vartheta}{\partial \Psi} = 0, \qquad \frac{\partial q}{\partial \Psi} - K(q) \frac{\partial \vartheta}{\partial \Phi} = 0. \tag{1.5}$$

Durch Einführen der TSCHAPLYGIN-Variablen beschränken wir uns auf Unterschall und daraus folgend auf elliptische Gleichungssysteme.

Wir schreiben statt (1.5)

$$\frac{\partial q}{\partial s} + \frac{\partial \vartheta}{\partial t} = 0, \qquad \frac{\partial q}{\partial t} - \frac{\partial \vartheta}{\partial s} = 0 \tag{1.6}$$

mit $s = s(\Phi, \Psi)$ und $t = t(\Phi, \Psi)$ als neuen Variablen, die sowohl das System (1.5) normalisieren, d. h. die elliptische Differentialform auf die kanonische zurückführen, als auch als neues Koordinatennetz verwendet werden.

s und t werden beim Übergang vom System (1.5) in das System (1.6) durch ein BELTRAMIsches Differentialgleichungspaar festgelegt:

$$\frac{\partial \Phi}{\partial s} - K \frac{\partial \Psi}{\partial t} = 0, \qquad \frac{\partial \Phi}{\partial t} + K \frac{\partial \Psi}{\partial s} = 0. \tag{1.7}$$

Die durch die Gleichung (1.1) bis (1.7) beschriebene Transformation kann auf jede beliebige Unterschall-Potentialströmung angewendet werden. Infolge der Linearität der Systeme (1.6) und (1.7) läßt sich ein ebenes Strömungsproblem in der s, t-Ebene wesentlich einfacher behandeln als in der Strömungsebene x, y. In der vorliegenden Arbeit soll deshalb auch die Behandlung von Profilgitterströmungen im wesentlichen in der s, t-Ebene beginnen.

Hat man aus dem System (1.7) die Potential- und Stromfunktion gefunden, wobei q und ϑ dem System (1.6) genügen müssen, so ergeben sich die Koordinaten x, y der Strömungsebene aus

$$dx + i\,dy = \frac{e^{i\vartheta}}{W} \left(d\Phi + i \frac{1}{\varrho} d\Psi \right). \tag{1.8}$$

Einer inkompressiblen als auch kompressiblen Strömung in der x, y-Ebene kann also mittels dieser Methode eine fiktive (inkompressible) Vergleichsströmung in der s, t-Ebene umkehrbar eindeutig zugeordnet werden.

Im Grenzfall der inkompressiblen Strömung in der x, y-Ebene ($M_b^2 \ll 1$) geht (1.3) über in

$$q = \ln W, \tag{1.9}$$

und K wird wegen $\varrho =$ konst. ebenfalls konstant, nämlich 1. Eine Lösung des BELTRAMIschen Differentialgleichungspaares (1.7) ist dann

$$\Phi(s, t) = s, \qquad \Psi(s, t) = t. \tag{1.10}$$

Die Potential- und Stromlinien der x, y-Ebene müssen also bei der Transformation in achsenparallele Geraden der s, t-Ebene übergehen. Für kompressible Strömung hingegen braucht bzw. darf diese Forderung nur näherungsweise erfüllt sein.

2. Aufstellung der Randbedingungen und Ableitung des Integralgleichungssystems

2.1. Darstellung des Gitters in der Vergleichsebene

In Bild 2.1 werden einige häufig wiederkehrende Begriffe, die bei Gitterströmungen verwendet werden, in Erinnerung gerufen bzw. neu definiert. Da es sich in dieser Arbeit um Gitterströmungen großer Umlenkung handelt, werden hier die Begriffe „innere“ und „äußere Profilstromlinie“ definiert. Während sich die innere Profilstromlinie aus dem vorderen „freien Anteil“ einer „Profilstromlinie“, der Saugseite des Profils und dem hinteren freien Anteil einer Profilstromlinie zusammensetzt, tritt bei der äußeren Profilstromlinie die Profildruckseite an die Stelle der Profilsaugseite.

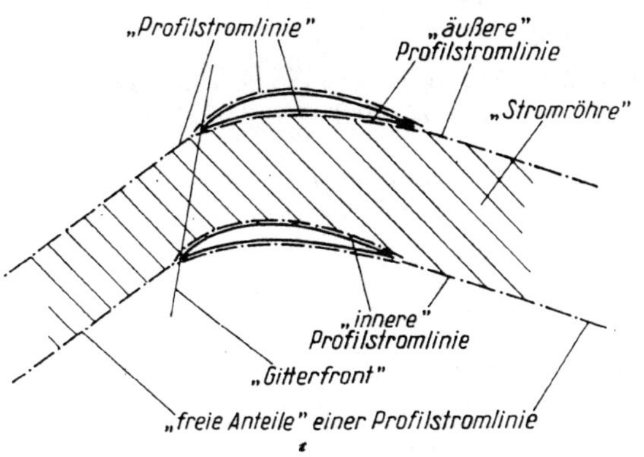

Bild 2.1

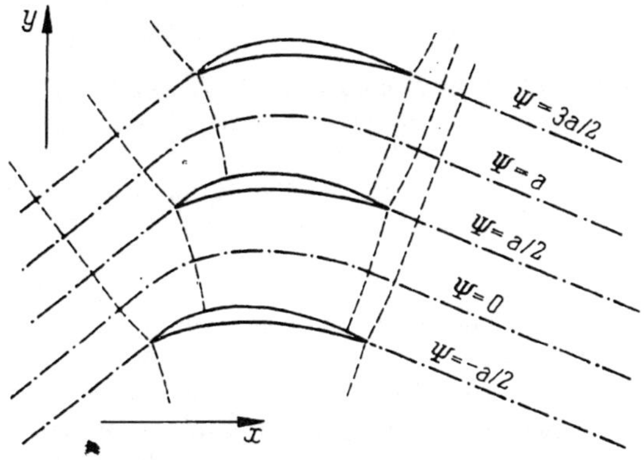

Bild 2.2. Darstellung eines Profilgitters in der x, y-Ebene

Zum leichteren Verständnis der Darstellung eines Gitters in der s, t-Ebene sei zunächst ein beliebiges, gerades und gestaffeltes Profilgitter in der x, y-Ebene dargestellt; siehe Bild 2.2. Unter dem Betrag von a ist in Bild 2.2 und im folgenden stets die in der Zeiteinheit durch eine Stromröhre der x, y-Ebene fließende *Menge* je Breiteneinheit zu verstehen. Wegen der Umlenkung der Strömung von $\vartheta_{-\infty}$ (Zuströmwinkel) auf $\vartheta_{+\infty}$ (Abströmwinkel) muß das Potential bei der Verfolgung einer Potentiallinie, auf der eine Profilkante liegen soll, zumindest an einer Profilkante springen. Eine in der gesamten Strömungsebene gültige stetige Bewertung der Potentiale ist also nicht möglich. Wir nehmen diese also zweckmäßigerweise für jede einzelne Stromröhre gesondert, und zwar derart vor, daß bei Parallelverschiebung von Stromröhren entlang der Gitterfront Potentiallinien mit gleichem Wert ineinander übergeführt werden können.

Wir sind nun in der Lage, eine der Darstellung in der x, y-Ebene entsprechende Darstellung eines Profilgitters in der s, t-Ebene zu zeichnen; siehe Bild 2.3. Einer Stromröhre der x, y-Ebene entspricht also jetzt ein zur s-Achse paralleler Streifen der Breite a, der sich von $s = -\infty$ bis $s = +\infty$ erstreckt. Die Profilstromlinien sind jetzt Geraden $t = \text{konst.} = (2\mu \pm 1)\, a/2$ mit $\mu = 0, \pm 1, \pm 2, \ldots$ Unter dem Betrag von a ist bei inkompressibler Strömung in der Strömungsebene x, y nunmehr in der Vergleichsebene s, t die Gitterteilung und konstante Stromröhrenbreite zu verstehen.

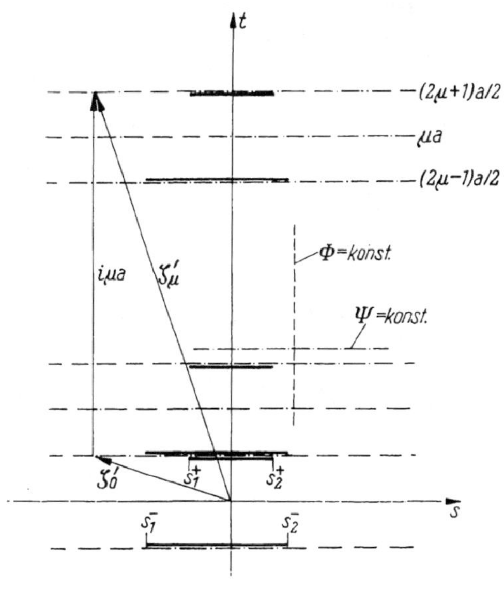

Bild 2.3. Darstellung eines Profilgitters in der s, t-Ebene

Innerhalb einer Stromröhre sind die Geschwindigkeitsfunktion q und der Strömungswinkel ϑ Funktionen von s und t:

$$q = q(s, t)\,, \qquad \vartheta = \vartheta(s, t)\,.$$

Für die innere und äußere *Profilstromlinie* hingegen vereinbaren wir folgende Abkürzungen:

$$\begin{aligned} q^+(s) &= q(s, +a/2)\,, & \vartheta^+(s) &= \vartheta(s, +a/2) & &\text{äußere Profilstromlinie,} \\ q^-(s) &= q(s, -a/2)\,, & \vartheta^-(s) &= \vartheta(s, -a/2) & &\text{innere Profilstromlinie,} \end{aligned} \tag{2.1}$$

zu deren Ermittlung wir die komplexe Funktion

$$g(\zeta) = q(\zeta) - i\,\vartheta(\zeta) \tag{2.2}$$

mit $\zeta = s + i\,t$ bilden und den Grenzübergang $t \to \pm a/2$ durchführen.

$g(\zeta)$ ist — unter der Voraussetzung von scharfen Profilkanten, stetiger Profilkontur und stoßfreiem Gitterein- und -austritt — im Gebiet der Stromröhre einschließlich dessen Rand regulär. Für $s \to -\infty$ und $s \to +\infty$ soll $g(\zeta)$ konstante (komplexe) Werte, nämlich die der Zu- und Abströmung annehmen:

$$g_{\mp\infty} = q_{\mp\infty} - i\,\vartheta_{\mp\infty}\,.$$

Die stark ausgezogenen Abschnitte der Geraden $t = (2\,\mu \pm 1)\,a/2$ von Bild 2.3 stellen die Gitterprofile dar. Auf ihnen soll $\vartheta = \vartheta^{+/-}(s)$ als Profilwinkelverteilung vorgegeben werden (kein wesentlicher Unterschied im Rechenablauf würde jedoch entstehen, sollte $q = q^{+/-}(s)$ als Profilgeschwindigkeitsverteilung vorgegeben sein). Wir vereinbaren im Zusammenhang damit folgende weitere Abkürzungen:

$$\begin{aligned} s_1^+ \leqq s \leqq s_2^+: &\quad \vartheta_p^+(s) = \vartheta(s, +a/2)\,, \quad q_p^+(s) = q(s, +a/2)\,,\\ s_1^- \leqq s \leqq s_2^-: &\quad \vartheta_p^-(s) = \vartheta(s, -a/2)\,, \quad q_p^-(s) = q(s, -a/2)\,, \end{aligned} \tag{2.3}$$

wobei Bild 2.3 Aufschluß über die Größen s_1^+, s_2^+, s_1^-, s_2^- liefert. Der stark ausgezogene Abschnitt $s_1^+ \leqq s \leqq s_2^+$ auf den Geraden $t = (2\,\mu + 1)\,a/2$ stellt also eine Profil*unter*seite dar, während $s_1^- \leqq s \leqq s_2^-$ auf $t = (2\,\mu - 1)\,a/2$ eine Profil*ober*seite darstellt.

$\vartheta_{-\infty}$ und $\vartheta_{+\infty}$ als Grenzwerte von $\vartheta(s, t)$ für $s \to \mp\infty$ sollen nicht vorgegeben werden. Eine Umströmung der Profilkanten kann somit vermieden werden. Als Bezugsgeschwindigkeitsfunktion möge der Ausdruck

$$q_b = (q_{-\infty} + q_{+\infty})/2$$

stehen und mit Null vorgegeben sein.

2.2. Aufstellung der Randbedingungen

Wir stellen nun jene Forderung auf, die im Grunde genommen eine Profilgitterströmung von einer Kanalströmung unterscheidet. Im gleichen Abstand vor den Profilvorderkanten bzw. im gleichen Abstand hinter den Profilhinterkanten müssen auf zwei beliebigen Profilstromlinien gleiche Geschwindigkeit und gleicher Strömungswinkel anzutreffen sein. Diese Forderung ist in ihrer vollen Gültigkeit in die s, t-Ebene zu übertragen. Gleicher Abstand in der x, y-Ebene heißt nämlich auch gleiche Potentialdifferenz in der x, y-Ebene, da selbstverständlich auf allen vorderen bzw. hinteren freien Anteilen der Profilstromlinien gleiche Geschwindigkeitsverteilung herrscht, und somit gleicher Abstand in der s, t-Ebene.

Die so in Worten formulierten Randbedingungen lauten mit

$$\Delta s_1 = |s_1^+ - s_1^-|\,, \qquad \Delta s_2 = |s_2^+ - s_2^-|$$

in Gleichungen:

$$\begin{aligned} s < s_1^-: &\quad q^-(s) = q^+(s + \Delta s_1)\,, \quad \vartheta^-(s) = \vartheta^+(s + \Delta s_1)\,,\\ s > s_2^-: &\quad q^-(s) = q^+(s - \Delta s_2)\,, \quad \vartheta^-(s) = \vartheta^+(s - \Delta s_2)\,. \end{aligned}$$

Bereits in Hinblick auf die spätere numerische Auswertung des zu gewinnenden Integralgleichungssystems schreiben wir aber statt der obigen zwei Gleichungssysteme:

$$\left.\begin{aligned} s < s_1^-: &\quad q^-(s) = q^+(s + \Delta s_1)\\ s < s_1^+: &\quad \vartheta^+(s) = \vartheta^-(s - \Delta s_1) \end{aligned}\right\} \text{vorderer freier Anteil der Profilstromlinie,}$$

$$\left.\begin{aligned} s > s_2^-: &\quad q^-(s) = q^+(s - \Delta s_2)\\ s > s_2^+: &\quad \vartheta^+(s) = \vartheta^-(s + \Delta s_2) \end{aligned}\right\} \text{hinterer freier Anteil der Profilstromlinie,} \tag{2.4}$$

wodurch ϑ auf die innere und q auf die äußere Profilstromlinie aufgeteilt werden.

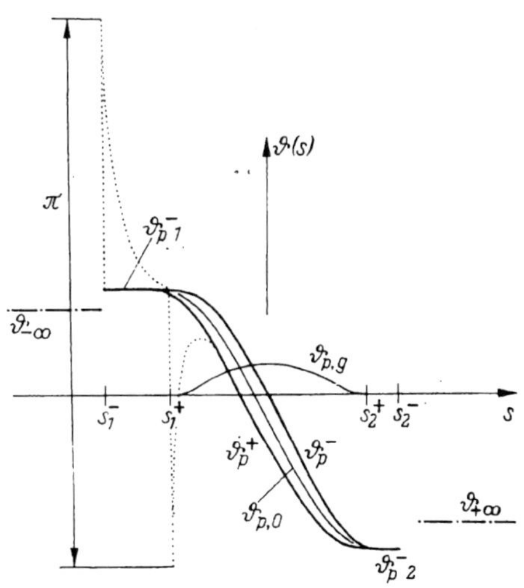

Bild 2.4

2.3. Vorgabe der Profilwinkelverteilung

Unter der Voraussetzung von scharfen Vorder- und Hinterkanten der Gitterprofile stellt sich die Profilwinkelverteilung etwa entsprechend der stark ausgezogenen Kurve von Bild 2.4 dar.

$\vartheta_p^+(s) = \vartheta_p^-(s) = \vartheta_{p,0}(s)$ wäre die Winkelverteilung einer beschleunigungs- und verzögerungsfreien Gitterströmung. Auf jenen Abschnitten der Profiloberseiten, von welchen ausgehend die Potentiallinien *nicht* auf eine gegenüberliegende Profilunterseite treffen, sondern in das Unendliche laufen ($s_1^- \leqq s \leqq s_1^+$ und $s_2^+ \leqq s \leqq s_2^-$), wurde im Rahmen dieser Arbeit ohne wesentliche Einschränkung der Allgemeinheit jeweils konstanter Strömungswinkel (ϑ_{p1}^- und ϑ_{p2}^-) angenommen.

Die Winkelverteilung einer beschleunigenden (mit vertauschtem Vorzeichen vor dem Winkel $\vartheta_{p,g}(s)$ ist es die einer verzögernden) Profilgitterströmung ist gemäß Bild 2.4

$$\vartheta_p^+(s) = \vartheta_{p,0}(s) - \vartheta_{p,g}(s)\,, \quad \vartheta_p^-(s) = \vartheta_{p,0}(s) + \vartheta_{p,g}(s)\,. \tag{2.5}$$

Punktiert ist in Bild 2.4 die Winkelverteilung eines vorn abgerundeten Profils eingetragen. Bei einer keilförmigen Profilspitze würde anstelle des Winkels π der entsprechend kleinere Öffnungswinkel des Keils stehen.

2.4. Ableitung des Integralgleichungssystems

Zur Überführung des in Abschnitt 2.2 formulierten Randwertproblems in ein System von zwei gekoppelten, singulären Integralgleichungen wenden wir die CAUCHYsche Integralformel an:

$$\mu \neq 0: \quad 0 = \frac{1}{2\pi i} \oint_{c_\mu} \frac{g(\zeta'_\mu)}{\zeta'_\mu - \zeta} d\zeta'_\mu, \qquad \mu = 0: \quad g(\zeta) = \frac{1}{2\pi i} \oint_{c_0} \frac{g(\zeta'_0)}{\zeta'_0 - \zeta} d\zeta'_0. \tag{2.6}$$

Wir beschränken damit die Lage des Aufpunktes $P(\zeta)$ auf die Grundstromröhre ($\mu = 0$). Sämtliche Rechnungen werden nämlich repräsentativ für das gesamte Gitter in der Grundstromröhre durchgeführt werden.

c_μ ist die Kontur der μ-ten Stromröhre. Sie unterscheidet sich von der der Grundstromröhre nur durch ihre entlang der t-Achse um die Strecke μa samt ihrer Belegung parallelverschobenen Lage. Es ist also

$$c_\mu = c_0\,; \qquad g(\zeta_\mu) = g(\zeta_0)$$

und gemäß Bild 2.3

$$\zeta_\mu = \zeta_0 + i\,\mu\,a\,.$$

Wir setzen dies in die Gleichungen (2.6) ein und bilden die Summe über alle μ, was einer Beachtung der Einflüsse aller Stromröhren mit dem Index $\mu \neq 0$ auf die Grundstromröhre $\mu = 0$ gleichkommt. Ausführen der Summation ergibt

$$g(\zeta) = \frac{1}{2\,a\,i} \oint_c g(\zeta') \coth\left[\frac{\pi}{a}(\zeta' - \zeta)\right] d\zeta'\,. \tag{2.7}$$

Der die Grundstromröhre betreffende Index Null wurde dabei bereits weggelassen.

Aufspalten von (2.7) in einen Real- und Imaginärteil ergibt mit den Abkürzungen

$$H_1(s, s', t) = \frac{\sinh\left[\frac{2\pi}{a}(s - s')\right]}{\cosh\left[\frac{2\pi}{a}(s - s')\right] + \cos\frac{2\pi}{a} t}, \qquad H_2(s, s', t) = \frac{1}{\cosh\left[\frac{2\pi}{a}(s - s')\right] + \cos\frac{2\pi}{a} t}$$

folgende Gleichungen für die Geschwindigkeitsfunktion $q(s, t)$ und den Strömungswinkel $\vartheta(s, t)$:

$$q(s, t) = \frac{1}{2a} \int_{-\infty}^{+\infty} H_1\,[\vartheta^+(s') - \vartheta^-(s')]\,ds' + \frac{1}{2a} \sin\frac{2\pi}{a} t \int_{-\infty}^{+\infty} H_2\,[q^+(s') - q^-(s')]\,ds'\,, \tag{2.8}$$

$$\vartheta(s, t) = \frac{1}{2}(\vartheta_{-\infty} + \vartheta_{+\infty}) - \frac{1}{2a} \int_{-\infty}^{+\infty} H_1\,[q^+(s') - q^-(s')]\,ds' + \frac{1}{2a} \sin\frac{2\pi}{a} t \int_{-\infty}^{+\infty} H_2\,[\vartheta^+(s') - \vartheta^-(s')]\,ds'\,. \tag{2.9}$$

Da wir speziell an der Geschwindigkeitsverteilung und den Strömungswinkeln auf der Profiloberfläche und den freien Anteilen der Profilstromlinien interessiert sind, führen wir in den Gleichungen (2.8) und (2.9) den Grenzübergang $t \to \pm\, a/2$ durch und erhalten:

$$q^+(s) + q^-(s) = -\frac{1}{a} \int_{-\infty}^{+\infty} \coth\left[\frac{\pi}{a}(s - s')\right] [\vartheta^+(s') - \vartheta^-(s')]\,ds'\,,$$

$$\vartheta^+(s) + \vartheta^-(s) = \vartheta_{-\infty} + \vartheta_{+\infty} + \frac{1}{a} \int_{-\infty}^{+\infty} \coth\left[\frac{\pi}{a}(s - s')\right] [q^+(s') - q^-(s')]\,ds'\,. \tag{2.10}$$

3. Lösung des Integralgleichungssystems

3.1. Überführung in ein lineares Gleichungssystem

Wir schreiben zunächst das System (2.10) wie folgt an:

$$q^+(i) + q^-(i) = -\frac{1}{a} \sum_{j=-N}^{+N} \int\limits_{\left(j-\frac{1}{2}\right)\Delta s}^{\left(j+\frac{1}{2}\right)\Delta s} \coth\left[\frac{\pi}{a}(i\,\Delta s - s')\right] [\vartheta^+(s') - \vartheta^-(s')]\, ds' , \tag{3.1}$$

$$\vartheta^+(i) + \vartheta^-(i) = \vartheta(-N) + \vartheta(+N) + \frac{1}{a} \sum_{j=-N}^{+N} \int\limits_{\left(j-\frac{1}{2}\right)\Delta s}^{\left(j+\frac{1}{2}\right)\Delta s} \coth\left[\frac{\pi}{a}(i\,\Delta s - s')\right] [q^+(s') - q^-(s')]\, ds'$$

mit $-N \leqq i \leqq N$.

Innerhalb der Intervalle $(i - (1/2))\,\Delta s \leqq s \leqq (i + (1/2))\,\Delta s$ sind die Funktionen q und ϑ näherungsweise als konstant anzunehmen und können somit vor das Integralzeichen gezogen werden. Übrig bleibt das Integral über die Kernfunktion:

$$\frac{1}{a} \int\limits_{\left(j-\frac{1}{2}\right)\Delta s}^{\left(j+\frac{1}{2}\right)\Delta s} \coth\left[\frac{\pi}{a}(i\,\Delta s - s')\right] ds' = C(i, j) = \frac{1}{\pi} \ln \left| \frac{\sinh\left[\frac{\pi\,\Delta s}{a}\left(i - j + \frac{1}{2}\right)\right]}{\sinh\left[\frac{\pi\,\Delta s}{a}\left(i - j - \frac{1}{2}\right)\right]} \right| . \tag{3.2}$$

Ein wesentlicher Vorteil dieser Lösungsmethode liegt offenbar darin, daß die singuläre Stelle $(i = j)$ keine besondere Behandlung erfordert:

$$C(i, i) = 0 ,$$

was im Fall des einfachen Ersetzens der Integrale durch Summen nicht der Fall wäre.

Die Funktionen $q(i)$ und $\vartheta(i)$ sind als Mittelwerte über die Intervallänge Δs zu verstehen. Dementsprechend ergibt sich die Profilwinkelverteilung, Gleichung (2.5), zu

$$\vartheta_p^{+/-}(i) = \frac{1}{\Delta s} \int\limits_{\left(i-\frac{1}{2}\right)\Delta s}^{\left(i+\frac{1}{2}\right)\Delta s} [\vartheta_{p,0}(s) \mp \vartheta_{p,g}(s)]\, ds . \tag{3.3}$$

Um den numerischen Aufwand in Grenzen zu halten, empfiehlt es sich, für $\vartheta_{p,0}(s)$ und $\vartheta_{p,g}(s)$ analytisch integrierbare — etwa harmonische — Funktionen zu wählen.

Es liegt am soeben entstehenden linearen Gleichungssystem zur Ermittlung von q und ϑ auf den Profilstromlinien, daß z. B. für $j \leqq -N - j_1^- + j_1^+$ bzw. $j \geqq N + j_2^+ - j_2^-$ (bezüglich $j_1^+, \ldots$ siehe die folgenden Randbedingungen) q^+ und ϑ^+ nicht gemeinsam konstant gesetzt werden können. Dadurch ist freilich eine Möglichkeit gegeben, die Faustregel zu überprüfen, daß etwa eine Gitterteilung vor den Profilvorderkanten bzw. hinter den Profilhinterkanten bereits die Zustände der Zu- bzw. Abströmung gelten. Es zeigt sich tatsächlich, daß die Ergebnisse von N praktisch nicht beeinflußt werden, falls

$$N \geqq \frac{1}{\Delta s} [\max(|s_1^-|, |s_2^-|) + a]$$

gewählt wird.

Die Randbedingungen (2.4) lauten damit in numerisch auswertbarer Form:

$$\begin{aligned}
j < j_1^+&: \quad q^+(j) \to q(j) , \\
j < j_1^-&: \quad q^-(j) \to q(j + j_1^- - j_1^+) , \\
j_1^+ \leqq j \leqq j_2^+&: \quad q^+(j) \to q_p^+(j) , \\
j_1^- \leqq j \leqq j_2^-&: \quad q^-(j) \to q_p^-(j) , \\
j > j_2^+&: \quad q^+(j) \to q(j) , \\
j > j_2^-&: \quad q^-(j) \to q(j + j_2^+ - j_2^-) ,
\end{aligned} \tag{3.4a}$$

$$
\left.
\begin{array}{lll}
j < j_1^+: & \vartheta^+(j) \to \vartheta(j + j_1^- - j_1^+)\,, & \\
j < j_1^-: & \vartheta^-(j) \to \vartheta(j)\,, & \\
j \leqq -N - j_1^- + j_1^+: & \vartheta^+(j) & \multirow{2}{*}{$\Big\}= \text{konst.} = \vartheta(-N)\,,$} \\
j \leqq -N: & \vartheta^-(j) & \\
j_1^+ \leqq j \leqq j_2^+: & \vartheta^+(j) \to \vartheta_p^+(j) \ \ldots \text{ vorgegeben}\,, & \\
j_1^- \leqq j \leqq j_2^-: & \vartheta^-(j) \to \vartheta_p^-(j) \ \ldots \text{ vorgegeben}\,, & \\
j > j_2^+: & \vartheta^+(j) \to \vartheta(j - j_2^+ + j_2^-)\,, & \\
j > j_2^-: & \vartheta^-(j) \to \vartheta(j)\,, & \\
j \geqq N + j_2^+ - j_2^-: & \vartheta^+(j) & \multirow{2}{*}{$\Big\}= \text{konst.} = \vartheta(N)\,.$} \\
j \geqq N: & \vartheta^-(j) &
\end{array}
\right. \tag{3.4b}
$$

Einsetzen in das System (3.1) liefert mit (3.2):

$$
\begin{aligned}
& q(i) + k_1^- q(i - j_1^- + j_1^+) + q_p^+(i) + q_p^-(i) + k_2^- q(i + j_2^+ - j_2^-) + \\
& \quad + \sum_{-N+1}^{-N-j_1^-+j_1^+} C(i,j)\,\vartheta(-N) - \sum_{-N+1}^{j_1^- -1} [C(i,j) - C(i, j - j_1^- + j_1^+)]\,\vartheta(j) - \\
& \quad - \sum_{j_2^- +1}^{N-1} [C(i,j) - C(i, j + j_2^+ - j_2^-)]\,\vartheta(j) + \sum_{N+j_2^+-j_2^-}^{N-1} C(i,j)\,\vartheta(N) \\
& \quad = \sum_{j_1^+}^{j_2^+} C(i,j)\,\vartheta_p^+(j) + \sum_{j_1^-}^{j_2^-} C(i,j)\,\vartheta_p^-(j)\,, \\
& \vartheta(i) + k_1^+ \vartheta(i + j_1^- - j_1^+) + k_2^+ \vartheta(i - j_2^+ + j_2^-) - \vartheta(-N) - \vartheta(N) - \\
& \quad - \sum_{-N}^{-N-j_1^-+j_1^+-1} C(i,j)\,q(j) - \sum_{-N-j_1^-+j_1^+}^{j_1^+-1} [C(i,j) - C(i, j + j_1^- - j_1^+)]\,q(j) - \\
& \quad - \sum_{j_2^+ +1}^{N+j_2^+-j_2^-} [C(i,j) - C(i, j - j_2^+ + j_2^-)]\,q(j) - \sum_{N+j_2^+-j_2^-+1}^{N} C(i,j)\,q(j) - \\
& \quad - \sum_{j_1^+}^{j_2^+} C(i,j)\,q_p^+(j) + \sum_{j_1^-}^{j_2^-} C(i,j)\,q_p^-(j) = -\vartheta_p^+(i) - \vartheta_p^-(i)
\end{aligned} \tag{3.5}
$$

mit $-N \leqq i \leqq +N$ und der folgenden Aufstellung für die vier Konstanten k_1^+, k_2^+, k_1^-, k_2^-:

	k_1^+	k_2^+
$j < j_1^+$	1	0
$j_1^+ \leqq j \leqq j_2^+$	0	0
$j_2^+ < j$	0	1

	k_1^-	k_2^-
$j < j_1^-$	1	0
$j_1^- \leqq j \leqq j_2^-$	0	0
$j_2^- < j$	0	1

Um nun das System (3.5) in *ein* lineares Gleichungssystem zu verwandeln, addieren wir zu allen Indizes i der zweiten Gleichung von (3.5) den konstanten Summanden $2N + 1$. Bezüglich des Index j ist eine detailliertere Aufschlüsselung notwendig:

$$
\begin{array}{lll}
-N \leqq j \leqq j_1^+ - 1: & q & \multirow{2}{*}{$\Big\}$ vorderer freier Anteil der Profilstromlinie,} \\
N + 1 \leqq j \leqq 2N + j_1^-: & \vartheta & \\
j_1^+ \leqq j \leqq j_2^+: & q_p^+\,, & \\
2N + j_1^- + 1 \leqq j \leqq 2N + j_2^- + 1: & q_p^-\,, & \\
j_2^+ \leqq j \leqq N: & q & \multirow{2}{*}{$\Big\}$ hinterer freier Anteil der Profilstromlinie.} \\
2N + j_2^- + 2 \leqq j \leqq 3N + 1: & \vartheta &
\end{array} \tag{3.6}
$$

3.2. Einige Bemerkungen zur Behandlung und Lösung des linearen Gleichungssystems

Die bisweilen auftretenden Schwierigkeiten (z. B.: Matrix A fast-singulär) bei der Lösung des linearen Gleichungssystems $A\,X = B$ machen eine Beschäftigung mit dieser im Prinzip bekannten Materie notwendig. Es sei allerdings nur auf die in diesem Zusammenhang wesentlichsten Punkte eingegangen.

Beginnend von links oben ($k = 1, 2, 3, \ldots$) werden im Rahmen einer GAUSS-Elimination immer kleinere quadratische Matrizen herangezogen. Während $R^{(k)}$ das absolut größte Element, also das „Pivotelement", in der „Restmatrix" $A^{(k)}$

darstellt, ist $H^{(k)}$ das *unter* der Hauptdiagonale absolut größte Element in der k-ten Spalte. Beim Übergang $A^{(k)} \to A^{(k+1)}$ dient die Ungleichung

$$R + (n - 1) H < 8 n R \tag{3.7}$$

mit n als dem Rang der Matrix, sozusagen als Weiche. Ist (3.7) erfüllt, so wird mit normaler Spaltenpivotsuche, andernfalls mit totaler Pivotsuche in $A^{(k+1)}$ fortgesetzt.

Eine Abschätzung, wie sie etwa Gleichung (3.7) verkörpert, nennt man „Pivotstrategie". Die vorliegende wurde aus der Erfahrung bei praktisch oft vorkommenden Systemen in der Fernmeldetechnik gewonnen, stammt von P. A. Businger [4], [5] und wurde in dieser Arbeit mit Erfolg auch auf Gitterströmungen angewendet. Bezüglich der in der überwiegenden Anzahl von Fällen notwendigen Nachiteration der Lösung siehe etwa [6].

4. Ermittlung der Koordinaten der Strömungsebene

4.1. Die Koordinaten der Profilstromlinie

Die Lösung von (1.8) lautet nach Aufspaltung in einen Real- und Imaginärteil und mit $t' = \pm a/2$, $dt' = 0$ für das Profil:

$$x_p^{+/-}(s) = \int_{s_1^{+/-}}^{s} \frac{\cos \vartheta_p^{+/-}(s')}{\exp q_p^{+/-}(s')}\, ds', \qquad y_p^{+/-}(s) = \int_{s_1^{+/-}}^{s} \frac{\sin \vartheta_p^{+/-}(s')}{\exp q_p^{+/-}(s')}\, ds' \tag{4.1}$$

sowie ebenfalls mit $t' = \pm a/2$, $dt' = 0$ für die freien Anteile der Profilstromlinien:

$$x^{+/-}(s) = \int_{s_1^{+/-}}^{s} \frac{\cos \vartheta^{+/-}(s')}{\exp q^{+/-}(s')}\, ds', \qquad y^{+/-}(s) = \int_{s_1^{+/-}}^{s} \frac{\sin \vartheta^{+/-}(s')}{\exp q^{+/-}(s')}\, ds'. \tag{4.2}$$

4.2. Erzeugung eines geschlossenen Strömungsprofils

Wir haben durch die Vorgabe der unteren Grenzen in den Integralen der Gleichungen (4.1) und (4.2) die Profilvorderkante in den Koordinatenursprung der x, y-Ebene gelegt. Die Hinterkante der Profil*unter*seite ist mit $s = s_2^+$ und die der Profil*ober*seite mit $s = s_2^-$ gegeben. Eine Bedingung dafür, daß wir ein technisch brauchbares mit einer sich an der Hinterkante schließenden Kontur erhalten, ist also:

$$\int_{s_1^+}^{s_2^+} \frac{\cos \vartheta_p^+(s')}{\exp q_p^+(s')}\, ds' = \int_{s_1^-}^{s_2^-} \frac{\cos \vartheta_p^-(s')}{\exp q_p^-(s')}\, ds', \qquad \int_{s_1^+}^{s_2^+} \frac{\sin \vartheta_p^+(s')}{\exp q_p^+(s')}\, ds' = \int_{s_1^-}^{s_2^-} \frac{\sin \vartheta_p^-(s')}{\exp q_p^-(s')}\, ds'$$

oder abgekürzt:

$$x_{p2}^+ = x_{p2}^-, \qquad y_{p2}^+ = y_{p2}^-. \tag{4.3}$$

Es muß nun gesagt werden, daß die vorliegende Betrachtungsweise (zum Unterschied zu den Singularitätenverfahren) weniger auf das Profil an sich, sondern vielmehr (wie bei den Kanaltheorien) auf die durch die innere und äußere Profilstromlinie gebildete Stromröhre abzielt. Die Profile selbst werden dabei von „Inseln" zwischen den einzelnen Stromröhren gebildet. Eine geschlossene Profilkontur würde also durch das lückenlose Aneinanderpassen der Stromröhren entlang der freien Anteile der Profilstromlinien gegeben sein.

Bei der zugrunde gelegten Betrachtungsweise, also der Betrachtung eines Einzelkanals, kommt das Problem des Zusammenpassens nicht zum Ausdruck und muß daher extra gestellt werden. Es ist dies auch der Grund dafür, daß die Bedingungen (4.3) nicht gleichzeitig mit der Ermittlung der x,y-Koordinaten erfüllt werden können. Mathematisch ausgedrückt: Das lineare Gleichungssystem (im Anschluß an das System (3.5)) wäre unter den Bedingungen (4.3) zu lösen, wobei aber in eben diesen Bedingungen die Lösung des linearen Gleichungssystems enthalten ist. Die letzte Formulierung deutet bereits auf jenes iterative Verfahren hin, das im folgenden abgeleitet werden soll und dessen Grenzwert ein geschlossenes Strömungsprofil sein wird. Wir verstehen zu diesem Zweck die Koordinaten der Profilober- und -unterseite mit den zwei Parametern f und δ:

$$x_p^+ = x_p^+(f, \delta)$$

mit analogen Beziehungen für die restlichen drei Koordinaten.

Die Korrekturfaktoren δ und f mögen multiplikativ einerseits zum Profilwinkel $\vartheta_{p,g}(s)$ (siehe Bild 2.4) und andererseits zur Durchflußmenge a treten, wobei aber zu beachten ist, daß die Änderung der Durchflußmenge nicht etwa auf Grund einer Änderung der Geschwindigkeit des Grundstroms erfolgen soll, sondern auf Grund einer Änderung der Gitterteilung.

Für zwei bestimmte Werte der Parameter stellen wir nun die Forderung nach einer geschlossenen Profilkontur auf:

$$x_{p2}^+(f, \delta) = x_{p2}^-(f, \delta), \qquad y_{p2}^+(f, \delta) = y_{p2}^-(f, \delta). \tag{4.4}$$

Das System (4.4) stellt zwei Gleichungen für die zwei Unbekannten f und δ dar. Es wird jedoch im Anschluß sofort klar, daß sie aus dem System (4.4) nur näherungsweise ausgerechnet werden können. Mehrmaliges aufeinanderfolgendes Ausrechnen der Parameter und Korrigieren der Durchflußmenge und der Profilwinkelverteilung, wobei sich f und δ, bzw. zumindest einer der beiden Faktoren Grenzwerten $\neq 1$ nähern, liefern jedoch eine der gewünschten Genauigkeit beliebig nahekommende Schließung des Strömungsprofils. In einem beliebigen Stadium der Iteration (z. B. Iterationsschritt i) gilt also

$$a_{i+1} = f_i\, a_i\,, \qquad \vartheta_{p,g}(s)_{i+1} = \delta_i\, \vartheta_{p,g}(s)_i\,. \tag{4.5}$$

Aufschluß über die Wirkungsweise des Faktors f erhalten wir aus der Potentialwirbelströmung. Diese gehorcht dem Gesetz

$$\frac{W^+}{W^-} = \frac{2\,r - b}{2\,r + b}\,. \tag{4.6}$$

Es sei dabei ein Kreisring von der Breite b und dem Mittelradius r ins Auge gefaßt (Die hochgestellten Indizes wurden in Anlehnung an eine Profilgitterströmung (+ ··· Profil*unter*seite, — ··· Profil*ober*seite) gesetzt.). Die zum Mittelradius r gehörige Geschwindigkeit sei mit $W = 1$ vorgegeben, womit b nicht nur die Kreisringbreite, sondern näherungsweise auch die durch den Kreisring pro Zeit- und Breiteneinheit fließende Menge darstellt.

Versehen wir nun die Kreisringbreite b bei festem r mit dem multiplikativen Faktor f, so ändern sich die Geschwindigkeiten $W^{+/-}$ bei einem beliebigen Iterationsschritt i wie folgt:

$$W_{i+1}^{+/-} = \frac{W_i^{+/-}}{(1 - f_i)\, W_i^{+/-} + f_i}\,. \tag{4.7}$$

Im gegenständlichen iterativen Verfahren sei angenommen, die Geschwindigkeiten im Gitter verhalten sich im Bereich der eigentlichen Umlenkung, also zwischen den Gitterprofilen, im Mittel dem Gesetz eines Potentialwirbels folgend. Für den Korrekturfaktor δ setzen wir die einfache Beziehung

$$\vartheta_p^{+/-}(s)_{i+1} = \vartheta_{p,0}(s) \mp \vartheta_{p,g}(s)_{i+1} = \vartheta_{p,0}(s) \mp \delta_i\, \vartheta_{p,g}(s)_i \tag{4.8}$$

an.

Der Ansatz (4.4) lautet demnach ausführlich:

$$\int_{s_1^+}^{s_2^+} [(1 - f)\, W^+(s') + f]\, \frac{\cos\,[\vartheta_{p,0}(s') - \delta\, \vartheta_{p,g}(s')]}{W^+(s')}\, ds' = \int_{s_1^-}^{s_2^-} [(1 - f)\, W^-(s') + f]\, \frac{\cos\,[\vartheta_{p,0}(s') + \delta\, \vartheta_{p,g}(s')]}{W^-(s')}\, ds'\,,$$

$$\int_{s_1^+}^{s_2^+} [(1 - f)\, W^+(s') + f]\, \frac{\sin\,[\vartheta_{p,0}(s') - \delta\, \vartheta_{p,g}(s')]}{W^+(s')}\, ds' = \int_{s_1^-}^{s_2^-} [(1 - f)\, W^-(s') + f]\, \frac{\sin\,[\vartheta_{p,0}(s') + \delta\, \vartheta_{p,g}(s')]}{W^-(s')}\, ds'$$

und daraus folgend näherungsweise:

$$\frac{(1 - f)\, W_m^+ + f}{(1 - f)\, W_m^- + f} = \frac{x_{p2}^- - y_{p2}^- \tan\,[(1 - \delta)\, \vartheta_{p,g}(0)]}{x_{p2}^+ + y_{p2}^+ \tan\,[(1 - \delta)\, \vartheta_{p,g}(0)]} = \frac{y_{p2}^- + x_{p2}^- \tan\,[(1 - \delta)\, \vartheta_{p,g}(0)]}{y_{p2}^+ - x_{p2}^+ \tan\,[(1 - \delta)\, \vartheta_{p,g}(0)]} \tag{4.9}$$

mit $W_m^{+/-}$ als den mittleren Geschwindigkeiten auf der Profilober/unterseite.

(4.9) stellt die zwei Gleichungen (4.4) für die zwei Unbekannten f und δ dar. Der zweite und dritte Ausdruck von (4.9) ergeben eine Gleichung für den Korrekturfaktor δ:

$$\delta = 1 - \frac{1}{\vartheta_{p,g}(0)} \arctan \left\{ \frac{x_{p2}^- y_{p2}^+ - x_{p2}^+ y_{p2}^-}{2\,(x_{p2}^+ x_{p2}^- + y_{p2}^+ y_{p2}^-)} \left[1 - \tan^2\left((1 - \delta)\, \vartheta_{p,g}(0)\right)\right] \right\}. \tag{4.10}$$

Aus dem ersten und zweiten Ausdruck von (4.9) kann sodann der Faktor f ermittelt werden. Mit der Hilfsgröße

$$\bar{f} = \frac{x_{p2}^- - y_{p2}^- \tan\,[(1 - \delta)\, \vartheta_{p,g}(0)]}{x_{p2}^+ + y_{p2}^+ \tan\,[(1 - \delta)\, \vartheta_{p,0}(0)]}$$

ergibt sich f schließlich zu

$$f = 1 \Big/ \left(1 - \frac{\bar{f} - 1}{\bar{f}\, W_m^- - W_m^+}\right). \tag{4.11}$$

Die Profilsehne S und der Abstand der Hinterkanten von Profilober- und -unterseite d sind gegeben mit

$$S = \frac{1}{2}\sqrt{(x_{p2}^+ + x_{p2}^-)^2 + (y_{p2}^+ + y_{p2}^-)^2}\,, \qquad d = \sqrt{(x_{p2}^+ - x_{p2}^-)^2 + (y_{p2}^+ - y_{p2}^-)^2}\,.$$

d ist also im Falle einer sich am Profilende nicht schließenden bzw. noch nicht geschlossenen Kontur jene Strecke, die zwischen der Hinterkante der Profiloberseite und der der Profilunterseite *ein und desselben* Profils liegt. Unter dem Ausdruck „Hinterkantenabstand" sei daher stets d und nicht etwa die Gitterteilung verstanden. Von einer geschlossenen Profilkontur wollen wir dann sprechen, bzw. die gewünschte Genauigkeit sei dann erfüllt, wenn mit einem vorgegebenen Faktor k_{sch} $d \leqq k_{sch} S$ gilt.

4.3. Staffelung und Teilung

Mit den Werten des Staffelungswinkels λ und der Teilung τ können wir das nunmehr an der Hinterkante geschlossene Strömungsprofil zu einem Gitter ergänzen. Aus der Kontinuitätsbedingung (siehe Bild 4.1) folgt:

$$\frac{a}{\tau} = W_{-\infty} \cos(\vartheta_{-\infty} + \lambda)$$
$$= W_{+\infty} \cos(\vartheta_{+\infty} + \lambda)\,. \tag{4.12}$$

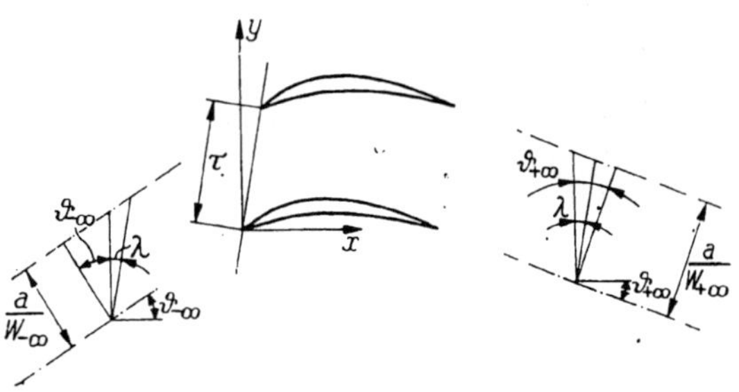

Bild 4.1

Der Staffelungswinkel λ ergibt sich daraus zu:

$$\lambda = \arctan \frac{W_{-\infty}\cos\vartheta_{-\infty} - W_{+\infty}\cos\vartheta_{+\infty}}{W_{-\infty}\sin\vartheta_{-\infty} - W_{+\infty}\sin\vartheta_{+\infty}}\,. \tag{4.13}$$

Durch die eingangs eingeführte Bezeichnung der Potentiallinien ist ein in der x,y-Ebene gegebenenfalls bestehender Staffelungswinkel in der Darstellung in der s,t-Ebene (siehe Bild 2.3) nicht ersichtlich und außerdem dort von keiner Bedeutung. Es sei jedoch darauf hingewiesen, daß der Staffelungswinkel λ aus (4.13) *nicht* der in der Technik gebräuchlichen Definition, nämlich $\lambda = \pi/2 - \sphericalangle$ (Gitterfront, Profilsehne) entspricht. Im Rahmen dieser Darstellung geben λ und die Gitterteilung τ (Gleichung (4.12)), letztere multipliziert mit dem laufenden Faktor $\mu = 0, \pm 1, \pm 2, \ldots$, die x, y-Koordinaten der Profilvorderkanten bei *beliebiger* Neigung der Profilsehne an.

5. Beispiel

Es wird ein Gitter behandelt mit ungefähr rechtwinkliger Umlenkung der Strömung, dessen Profile als „dick" bezeichnet werden können.

Die Profilwinkelverteilung wurde mit der Abkürzung

$$b = -s_1^+ + s_2^+ = \Delta s\,(-j_1^+ + j_2^+ + 1)$$

und Gleichung (2.5) folgend angenommen zu:

$$\vartheta_p^{+/-}(s) = \frac{1}{2}\left[\vartheta_{p1}^- + \vartheta_{p2}^- - (\vartheta_{p1}^- - \vartheta_{p2}^-)\sin\frac{\pi}{b}s \mp \vartheta_{p,g}(0)\left(1 + \cos\frac{2\pi}{b}s\right)\right].$$

Wenden wir darauf (3.4) an, so erhalten wir eine für die numerische Auswertung bequem verwendbare Form der Profilwinkelverteilung:

$$\vartheta_p^{+/-}(i) = \frac{1}{2}\left[\vartheta_{p1}^- + \vartheta_{p2}^- - (\vartheta_{p1}^- - \vartheta_{p2}^-)\frac{2b}{\pi\Delta s}\sin\frac{\pi\Delta s}{2b}\sin\frac{\pi}{b}i \mp \vartheta_{p,g}(0)\left(1 + \frac{b}{\pi\Delta s}\sin\frac{\pi\Delta s}{b}\cos\frac{2\pi}{b}i\right)\right]. \tag{5.1}$$

Vorgegeben sind:

$N = 18\,,\quad j_1^- = -7\,,\quad -j_1^+ = j_2^+ = 4\,,\quad j_2^- = 8\,,$

$\Delta s = 1.0\,,\quad \vartheta_{p1}^- = 0.65\,,\quad \vartheta_{p2}^- = -1.0\,,$

$\left.\begin{array}{l} a = 4.0\,, \\ \vartheta_{p,g}(0) = 0.1 \end{array}\right\}$ Anfangswerte der Iteration $d \to 0$.

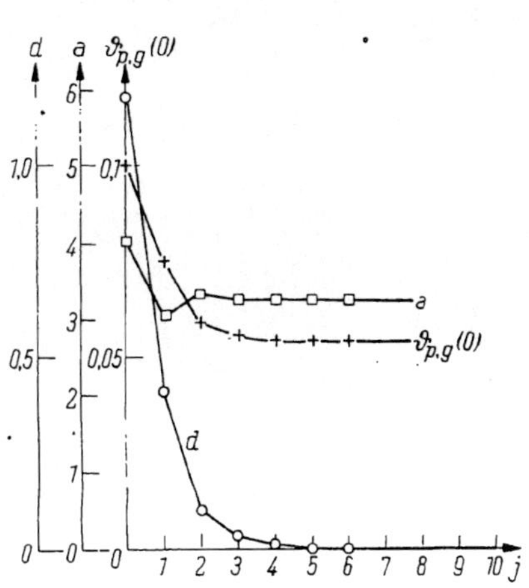

Bild 5.1

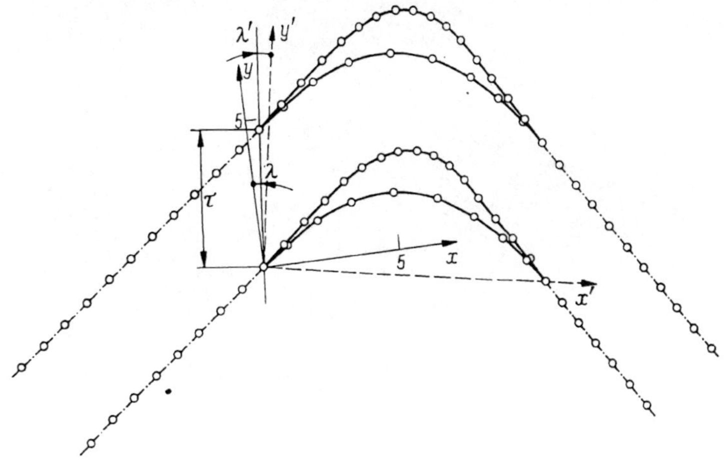

Bild 5.2

Bild 5.1 zeigt die Änderung der durch eine Stromröhre fließenden Menge a, also der Stromröhrenbreite in der Vergleichsebene s, t, und des Profilwinkels $\vartheta_{p,g}(0)$ (siehe Bild 2.4) im Laufe der Iteration d (Hinterkantenabstand) gegen Null. Die Endwerte der Iteration $d \to 0$ sind

$$a = 3.2317\,, \qquad \vartheta_{p,g}(0) = 0.05366\,.$$

Bild 5.2 zeigt das Gitter in der Strömungsebene x, y. λ' ist der in der Technik gebräuchliche Gitterstaffelungswinkel, x', y' sind die dazugehörigen Koordinatenrichtungen.

Die über x'' (senkrecht auf die Gitterfront) aufgetragene Profilgeschwindigkeitsverteilung zeigt Bild 5.3 (stark ausgezogener Kurvenzug). Der schwach ausgezogene Kurvenzug bildet die Kontrolle der Rechnung durch eine Messung nach der Elektroanalogie (siehe [7]).

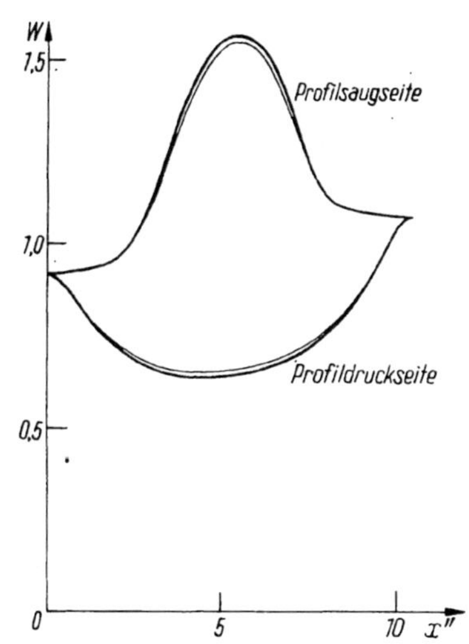

Bild 5.3

Literatur

1 GRETLER, W., Eine indirekte Methode zur Berechnung der ebenen Unterschallströmung, Journal de Mécanique 7, S. 83–96 (1966).
2 GRETLER, W., Erweiterung der inversen Methode zur Ermittlung ebener Unterschallströmungen auf den Fall der Umströmung angestellter und gewölbter Profile, Journal de Mécanique 8, S. 211–228 (1969).
3 CHRISTIANOWITSCH, S. A., Die Gasströmung um einen Körper bei hohen Unterschallgeschwindigkeiten, Arbeiten des Zentralinstitutes für Aero-Hydromechanik, Moskau 1940.
4 BUSINGER, P. A., MIDAS-Solution of Linear Algebraic Equations, Numerical Mathematics Program Library Project, Bell Telephone Laboratories, Inc., Murray Hill, N. J., USA (1970).
5 BUSINGER, P. A., Monitoring the Numerical Stability of Gaussian Elimination, Numer. Math. 16 (1971).
6 ISAACSON, E., and KELLER, H. B., Analysis of Numerical Methods, New York, London, Sydney: John Wiley & Sons, Inc., 1966.
7 HACKESCHMIDT, M., Grundlagen der Strömungstechnik II, Leipzig: Deutscher Verlag für Grundstoffindustrie, 1970.

Eingereicht am 15. 3. 1973

Anschrift: Dipl.-Ing. Dr. techn. RUDOLF M. MARKYTAN, 115 Suellen Road, Apt. B-3, Schenectady, New York 12309, U. S. A.

ZAMM 54, 103 – 118 (1974)

H. BUFLER

Elastisches Mehrschichtsystem unter asymmetrischer Belastung*)

Es wird ein systematischer Lösungsweg für den räumlichen Spannungs- und Verschiebungszustand eines aus elastisch-isotropen, planparallelen Schichten bestehenden Verbundsystems in Zylinderkoordinaten entwickelt. Die Anwendung von Fourier- und Hankeltransformationen in Verbindung mit einer konsequenten matrizenmäßigen Aufbereitung erlaubt eine übersichtliche Darstellung der Ergebnisse.

Author develops a systematical method of solution for the spatial state of stress and displacement of an elastic-isotropic layered medium using cylindrical coordinates. Fourier- and Hankeltransforms as well as the consequent application of matrix analysis give clear arranged results.

Представлен систематический путь решения пространственного состояния напряжения и перемещения в цилиндрических координатах для системы, состоящей из упругих, изотропных, плоскопараллельных и связанных между собой слоев. Путём применения перестановок Фурье и Ганкеля совместно с последовательной матричной обработкой представляется возможным особенно наглядно представить результаты.

1. Einleitung

Es wird ein aus beliebig vielen planparallelen, elastisch-isotropen Schichten bestehendes Verbundsystem zugrunde gelegt. Die seitlichen Abmessungen seien so groß, daß man von unendlich ausgedehnten Schichten sprechen kann. Eine Schicht k wird dann allein durch ihre Dicke h_k sowie ihre Elastizitätskonstanten E_k (Elastizitätsmodul) und ν_k (Querdehnzahl) charakterisiert. Zwischen den Schichten bestehe entweder vollkommener Verbund oder keinerlei Haftung; im zweiten Fall muß die Belastung so sein, daß kein Abheben der Schichten voneinander eintreten kann und das Gleichgewicht ohne Übertragung von Schubspannungen möglich ist. Zu bestimmen ist der vollständige Spannungs- und Verschiebungszustand infolge vorgeschriebener Randkräfte bzw. Randverschiebungen an den Deckflächen des Verbundsystems.

Diese räumliche Aufgabe wurde bei Verwendung kartesischer Koordinaten und der zweidimensionalen FOURIERtransformation in [1] einer systematischen Lösung zugeführt. Im vorliegenden Beitrag soll eine entsprechende Lösung in Zylinderkoordinaten entwickelt werden, die insbesondere dann von Vorteil ist, wenn die Belastung auf einer Kreisfläche vorgeschrieben ist. KUO [2] befaßte sich zwar mit derselben Aufgabe, doch fehlt bei ihm eine systematische Aufbereitung der Gleichungen unter Ausnutzung der speziellen Matrizeneigenschaften, so daß dort sowohl die formelmäßige als auch die numerische Auswertung zu Schwierigkeiten führen kann.

Die partiellen Ausgangs-Differentialgleichungen werden im folgenden durch Anwendung der endlichen FOURIER-Transformation bezüglich der Winkelkoordinate φ sowie der HANKEL-Transformation bezüglich der Radialkoordinate r auf gewöhnliche Differentialgleichungen bezüglich der Koordinate z (senkrecht zur Schichtebene) reduziert; ihre Integration führt unmittelbar zur Übertragungsmatrix der Schicht k. Aus den Rand- und Übergangsbedingungen erhält man ein lineares inhomogenes Gleichungssystem für die Transformierten der Komponenten des Spannungs- und Verschiebungsvektors an den Außen- und Innenrändern; mit ihnen steht der gesamte Spannungs- und Verschiebungszustand fest.

Im Falle der Axialsymmetrie ergibt sich eine Entkopplung: Das Problem 6. Ordnung zerfällt in ein Problem 4. Ordnung (axialsymmetrischer torsionsfreier Verzerrungszustand) und in ein Problem 2. Ordnung (axialsymmetrische Torsion). Dieser Sonderfall wurde in [3] für querisotropes Material gemeinsam mit einem weiteren zweidimensionalen Verzerrungszustand behandelt, wobei sich eine vollständige Analogie herausstellte.

Das axialsymmetrische Problem einer Einzelschicht findet man ausführlich diskutiert bei SNEDDON [4]; die Verallgemeinerung auf den Fall der asymmetrischen Belastung wurde durch MUKI [5] vorgenommen.

Weitere umfangreiche Literaturangaben über Schichtkörper sind in der kürzlich herausgekommenen Bibliographie [6] enthalten.

2. Matrizenmäßige Formulierung der Ausgangs-Differentialgleichungen

Die Gleichungen (2.1)–(2.6) repräsentieren das HOOKEsche Gesetz für Isotropie; die Verzerrungskomponenten sind dort durch die Verschiebungsableitungen ausgedrückt. Die Gleichgewichtsbedingungen bei Abwesenheit von Volumkräften finden sich in (2.7)–(2.9). Die Verschiebungskomponenten in Richtung von r, φ und z sind u_r, u_φ, u_z, die Spannungskomponenten werden mit σ_{rr}, $\sigma_{\varphi\varphi}$, σ_{zz}, $\sigma_{r\varphi}$, σ_{zr}, $\sigma_{z\varphi}$ bezeichnet.

*) Auszug aus dem im Nov. 1970 in Piešťany/CSSR auf dem Symposium „Kontaktaufgaben der Baukonstruktionen und der elastischen Unterlage" vom Verf. gehaltenen Vortrag über „Rationelle und exakte Berechnung des Spannungs- und Verschiebungszustandes eines elastischen Mehrschichtsystems".

$$\frac{\partial u_r}{\partial r} = \frac{1}{E}(\sigma_{rr} - \nu\,\sigma_{\varphi\varphi} - \nu\,\sigma_{zz}), \qquad (2.1)$$

$$\frac{\partial u_\varphi}{r\,\partial\varphi} + \frac{u_r}{r} = \frac{1}{E}(\sigma_{\varphi\varphi} - \nu\,\sigma_{zz} - \nu\,\sigma_{rr}), \qquad (2.2)$$

$$\frac{\partial u_z}{\partial z} = \frac{1}{E}(\sigma_{zz} - \nu\,\sigma_{rr} - \nu\,\sigma_{\varphi\varphi}); \qquad (2.3)$$

$$\frac{\partial u_r}{r\,\partial\varphi} + \frac{\partial u_\varphi}{\partial r} - \frac{u_\varphi}{r} = \frac{2(1+\nu)}{E}\sigma_{r\varphi}, \qquad (2.4)$$

$$\frac{\partial u_\varphi}{\partial z} + \frac{\partial u_z}{r\,\partial\varphi} = \frac{2(1+\nu)}{E}\sigma_{\varphi z}, \qquad (2.5)$$

$$\frac{\partial u_z}{\partial r} + \frac{\partial u_r}{\partial z} = \frac{2(1+\nu)}{E}\sigma_{zr}, \qquad (2.6)$$

$$\frac{\partial \sigma_{rr}}{\partial r} + \frac{\partial \sigma_{\varphi r}}{r\,\partial\varphi} + \frac{\partial \sigma_{zr}}{\partial z} + \frac{1}{r}(\sigma_{rr} - \sigma_{\varphi\varphi}) = 0, \qquad (2.7)$$

$$\frac{\partial \sigma_{r\varphi}}{\partial r} + \frac{2}{r}\sigma_{r\varphi} + \frac{\partial \sigma_{\varphi\varphi}}{r\,\partial\varphi} + \frac{\partial \sigma_{z\varphi}}{\partial z} = 0, \qquad (2.8)$$

$$\frac{\partial \sigma_{rz}}{\partial r} + \frac{1}{r}\sigma_{rz} + \frac{\partial \sigma_{\varphi z}}{r\,\partial\varphi} + \frac{\partial \sigma_{zz}}{\partial z} = 0. \qquad (2.9)$$

Auflösung von (2.1), (2.2) und (2.4) nach σ_{rr}, $\sigma_{\varphi\varphi}$ und $\sigma_{r\varphi}$ ergibt

$$\sigma_{rr} = \frac{\nu}{1-\nu}\sigma_{zz} + \frac{E}{1-\nu^2}\left[\frac{\partial u_r}{\partial r} + \nu\left(\frac{u_r}{r} + \frac{\partial u_\varphi}{r\,\partial\varphi}\right)\right], \qquad (2.10)$$

$$\sigma_{\varphi\varphi} = \frac{\nu}{1-\nu}\sigma_{zz} + \frac{E}{1-\nu^2}\left[\nu\frac{\partial u_r}{\partial r} + \frac{u_r}{r} + \frac{\partial u_\varphi}{r\,\partial\varphi}\right], \qquad (2.11)$$

$$\sigma_{r\varphi} = \frac{E}{2(1+\nu)}\left[\frac{\partial u_r}{r\,\partial\varphi} + \frac{\partial u_\varphi}{\partial r} - \frac{u_\varphi}{r}\right]. \qquad (2.12)$$

Eliminiert man diese Größen in den übrigen Gleichungen, so bekommt man nach geeigneter Zusammenfassung mit dem „Zustandsvektor"

$$\boldsymbol{a} = \{\sigma_{zz}, \sigma_{zr}, \sigma_{z\varphi}, u_\varphi^*, u_r^*, u_z^*\}^T \qquad (2.13)$$

und der Differentialmatrix $\boldsymbol{A}$ gemäß (2.14) folgende Matrizen-Differentialgleichung:

$$\frac{\partial \boldsymbol{a}}{\partial z} = \boldsymbol{A}\,\boldsymbol{a}. \qquad (2.15)$$

Aus Dimensionsgründen wurden hier die den Spannungen dimensionsgleichen Verschiebungen

$$u_i^* = \frac{E^*}{h^*}u_i \qquad (i = r, \varphi, z) \qquad (2.16)$$

mit E^* als Bezugselastizitätsmodul und h^* als Bezugsdicke eingeführt. Ferner lassen sich die Gleichungen für die restlichen Spannungskomponenten (2.10)–(2.12) mit dem Vektor (Spaltenmatrix)

$$\boldsymbol{b} = \{\sigma_{rr} + \sigma_{\varphi\varphi}, \sigma_{rr} - \sigma_{\varphi\varphi}, 2\,\sigma_{r\varphi}\}^T \qquad (2.17)$$

und der Differentialmatrix

$$\boldsymbol{A} = \begin{bmatrix} 0 & -\left(\frac{\partial}{\partial r} + \frac{1}{r}\right) & -\frac{\partial}{r\,\partial\varphi} & 0 & 0 & 0 \\ -\frac{\nu}{1-\nu}\frac{\partial}{\partial r} & 0 & 0 & +\frac{h^*}{2(1-\nu^2)}\frac{E}{E^*}\left[\frac{3-\nu}{r^2}\frac{\partial}{\partial\varphi} - \frac{1+\nu}{r}\frac{\partial^2}{\partial r\,\partial\varphi}\right] & -\frac{h^*}{1-\nu^2}\frac{E}{E^*}\left[\frac{\partial^2}{\partial r^2} + \frac{1}{r}\frac{\partial}{\partial r} - \frac{1}{r^2} + \frac{1-\nu}{2}\frac{\partial^2}{r^2\,\partial\varphi^2}\right] & 0 \\ -\frac{\nu}{1-\nu}\frac{\partial}{r\,\partial\varphi} & 0 & 0 & -\frac{h^*}{1-\nu^2}\frac{E}{E^*}\left[\frac{\partial^2}{r^2\,\partial\varphi^2} + \frac{1-\nu}{2}\left(\frac{\partial^2}{\partial r^2} + \frac{1}{r}\frac{\partial}{\partial r} - \frac{1}{r^2}\right)\right] & -\frac{h^*}{2(1-\nu^2)}\frac{E}{E^*}\left[\frac{3-\nu}{r^2}\frac{\partial}{\partial\varphi} + \frac{1+\nu}{r}\frac{\partial^2}{\partial r\,\partial\varphi}\right] & 0 \\ 0 & 0 & \frac{2(1+\nu)}{h^*}\frac{E^*}{E} & 0 & 0 & -\frac{\partial}{r\,\partial\varphi} \\ 0 & \frac{2(1+\nu)}{h^*}\frac{E^*}{E} & 0 & 0 & 0 & -\frac{\partial}{\partial r} \\ \frac{(1+\nu)(1-2\nu)}{(1-\nu)h^*}\frac{E^*}{E} & 0 & 0 & -\frac{\nu}{1-\nu}\frac{\partial}{r\,\partial\varphi} & -\frac{\nu}{1-\nu}\left(\frac{\partial}{\partial r} + \frac{1}{r}\right) & 0 \end{bmatrix}. \qquad (2.14)$$

$$B = \begin{bmatrix} \frac{2\nu}{1-\nu} & 0 & 0 & \frac{h^*}{1-\nu}\frac{E}{E^*}\frac{\partial}{r\,\partial\varphi} & \frac{h^*}{1-\nu}\frac{E}{E^*}\left(\frac{\partial}{\partial r}+\frac{1}{r}\right) & 0 \\ 0 & 0 & 0 & \frac{-h^*}{1+\nu}\frac{E}{E^*}\frac{\partial}{r\,\partial\varphi} & \frac{h^*}{1+\nu}\frac{E}{E^*}\left(\frac{\partial}{\partial r}-\frac{1}{r}\right) & 0 \\ 0 & 0 & 0 & \frac{h^*}{1+\nu}\frac{E}{E^*}\left(\frac{\partial}{\partial r}-\frac{1}{r}\right) & \frac{h^*}{1+\nu}\frac{E}{E^*}\frac{\partial}{r\,\partial\varphi} & 0 \end{bmatrix} \tag{2.18}$$

in folgender Form schreiben:

$$b = B\,a\,. \tag{2.19}$$

3. Lösung für die Einzelschicht

3.1. Reduktion auf gewöhnliche Differentialgleichungen

a) Erster Reduktionsschritt: Anwendung der endlichen FOURIER-Transformation

Sie ist in komplexer Form definiert durch [4]

$$\hat{f}(n) = \mathcal{F}_0\{f(\varphi)\} = \frac{1}{\sqrt{2\pi}} \int_{-\pi}^{+\pi} f(\varphi)\, e^{in\varphi}\, d\varphi\,, \qquad \text{bzw.} \qquad \hat{\hat{f}}(n) = \mathcal{F}_1\{f(\varphi)\} = \frac{1}{j}\, \mathcal{F}_0\{f(\varphi)\} \tag{3.1}$$

mit

$$j = i\,\frac{n}{|n|}\,, \qquad i = \sqrt{-1}\,, \qquad (n \text{ ganze Zahl}) \qquad \text{und} \qquad j = i \qquad \text{für} \quad n = 0\,.$$

Die inversen Transformierten sind

$$f(\varphi) = \left\langle \begin{matrix} \mathcal{F}_0^{-1}\{\hat{f}(n)\} \\ \mathcal{F}_1^{-1}\{\hat{\hat{f}}(n)\} \end{matrix} \right\rangle = \frac{1}{\sqrt{2\pi}} \sum_{n=-\infty}^{\infty} \left\langle \begin{matrix} \hat{f}(n) \\ j\,\hat{\hat{f}}(n) \end{matrix} \right\rangle e^{-in\varphi}\,, \tag{3.2}$$

falls $f(\varphi)$ die DIRICHLETschen Bedingungen im Intervall $-\pi < \varphi < +\pi$ erfüllt und falls das Integral $\int_{-\pi}^{+\pi} f(\varphi)\, d\varphi$ absolut konvergiert.

An einer Sprungstelle erhält man $\frac{1}{2}\,[f(\varphi+0) + f(\varphi-0)]$ anstelle von $f(\varphi)$. Mittels partieller Integration kann man leicht die Zusammenhänge

$$\frac{d\hat{f}}{d\varphi} = -\,i\,n\,\hat{f}\,, \qquad \text{falls } f(\pi) = f(-\pi) \text{ ist, und}$$

$$\frac{d^2\hat{f}}{d\varphi^2} = -\,n^2\,\hat{f}\,, \qquad \text{falls außerdem } \frac{df}{d\varphi}(\pi) = \frac{df}{d\varphi}(-\pi) \text{ ist,}$$

nachweisen. Folglich gilt

$$\begin{aligned} \frac{d\hat{f}}{d\varphi} &= |n|\,\hat{\hat{f}}\,; & \frac{d\hat{\hat{f}}}{d\varphi} &= -\,|n|\,\hat{f}\,; \\ \frac{d^2\hat{f}}{d\varphi^2} &= -\,n^2\,\hat{f}\,; & \frac{d^2\hat{\hat{f}}}{d\varphi^2} &= -\,n^2\,\hat{\hat{f}}\,. \end{aligned} \tag{3.3}$$

Die Anwendung der Operationen

$$F = \operatorname{diag}(\mathcal{F}_0, \mathcal{F}_0, \mathcal{F}_1, \mathcal{F}_1, \mathcal{F}_0, -\mathcal{F}_0) \qquad \text{bzw.} \qquad F' = \operatorname{diag}(\mathcal{F}_0, \mathcal{F}_0, \mathcal{F}_1) \tag{3.4), (3.5}$$

auf Gl. (2.15) bzw. Gl. (2.19) liefert sodann

$$\frac{\partial}{\partial z}(F\,a) = F\,A\,a\,,$$

$$F'\,b = F'\,B\,a$$

und bei Beachtung von (3.3)

$$\frac{\partial}{\partial z}\hat{a} = \hat{A}\,\hat{a}\,, \tag{3.6}$$

$$\hat{b} = \hat{B}\,\hat{a} \tag{3.7}$$

mit den transformierten Spaltenvektoren

$$\hat{a} = F\,a = \{\hat{\sigma}_{zz}, \hat{\sigma}_{zr}, \hat{\hat{\sigma}}_{z\varphi}, \hat{\hat{u}}_\varphi^*, \hat{u}_r^*, -\,\hat{u}_z^*\}^T\,, \tag{3.8}$$

$$\hat{b} = F'\,b = \{(\sigma_{rr} + \sigma_{\varphi\varphi})^\wedge, (\sigma_{rr} - \sigma_{\varphi\varphi})^\wedge, 2\,\hat{\hat{\sigma}}_{r\varphi}\}^T\,, \tag{3.9}$$

sowie den neuen Differentialmatrizen

$$\hat{A} = \begin{bmatrix} 0 & -\left(\frac{\partial}{\partial r}+\frac{1}{r}\right) & -\frac{|n|}{r} & 0 & 0 & 0 \\ -\frac{\nu}{1-\nu}\frac{\partial}{\partial r} & 0 & 0 & \frac{h^*}{2(1-\nu^2)}\frac{E}{E^*}\left[(3-\nu)\frac{|n|}{r^2}-(1+\nu)\frac{|n|}{r}\frac{\partial}{\partial r}\right] & \frac{-h^*}{1-\nu^2}\frac{E}{E^*}\left[\frac{\partial^2}{\partial r^2}+\frac{1}{r}\frac{\partial}{\partial r}-\frac{1}{r^2}-\frac{1-\nu}{2}\frac{n^2}{r^2}\right] & 0 \\ \frac{\nu}{1-\nu}\frac{|n|}{r} & 0 & 0 & \frac{-h^*}{1-\nu^2}\frac{E}{E^*}\left[-\frac{n^2}{r^2}+\frac{1-\nu}{2}\left(\frac{\partial^2}{\partial r^2}+\frac{1}{r}\frac{\partial}{\partial r}-\frac{1}{r^2}\right)\right] & \frac{h^*}{2(1-\nu)^2}\frac{E}{E^*}\left[(3-\nu)\frac{|n|}{r^2}+(1+\nu)\frac{|n|}{r}\frac{\partial}{\partial r}\right] & 0 \\ 0 & 0 & \frac{2(1+\nu)}{h^*}\frac{E^*}{E} & 0 & 0 & -\frac{|n|}{r} \\ 0 & \frac{2(1+\nu)}{h^*}\frac{E^*}{E} & 0 & 0 & 0 & \frac{\partial}{\partial r} \\ -\frac{(1+\nu)(1-2\nu)}{(1-\nu)h^*}\frac{E}{E^*} & 0 & 0 & \frac{\nu}{1-\nu}\frac{|n|}{r} & \frac{\nu}{1-\nu}\left(\frac{\partial}{\partial r}+\frac{1}{r}\right) & 0 \end{bmatrix} \tag{3.10}$$

$$\tilde{A} = \begin{bmatrix} 0 & -\frac{1}{2}\left(\frac{\partial}{\partial r}+\frac{1+|n|}{r}\right) & -\frac{1}{2}\left(\frac{\partial}{\partial r}+\frac{1-|n|}{r}\right) & 0 & 0 & 0 \\ \frac{-\nu}{1-\nu}\left(\frac{\partial}{\partial r}-\frac{|n|}{r}\right) & 0 & 0 & -\frac{h^*}{4(1-\nu)}\frac{E}{E^*}\left[\frac{\partial^2}{\partial r^2}-\frac{1}{r^2}(1-n^2)+\frac{1-2|n|}{r}\frac{\partial}{\partial r}\right] & \frac{-h^*(3-\nu)}{4(1-\nu^2)}\frac{E}{E^*}\left[\frac{\partial^2}{\partial r^2}-\frac{(1+|n|)^2}{r^2}+\frac{1}{r}\frac{\partial}{\partial r}\right] & 0 \\ \frac{-\nu}{1-\nu}\left(\frac{\partial}{\partial r}+\frac{|n|}{r}\right) & 0 & 0 & \frac{-h^*(3-\nu)}{4(1-\nu^2)}\frac{E}{E^*}\left[\frac{\partial^2}{\partial r^2}-\frac{(1-|n|)^2}{r^2}+\frac{1}{r}\frac{\partial}{\partial r}\right] & -\frac{h^*}{4(1-\nu)}\frac{E}{E^*}\left[\frac{\partial^2}{\partial r^2}-\frac{1-n^2}{r^2}+\frac{1+2|n|}{r}\frac{\partial}{\partial r}\right] & 0 \\ 0 & 0 & \frac{2(1+\nu)}{h^*}\frac{E^*}{E} & 0 & 0 & \frac{1}{2}\left(\frac{\partial}{\partial r}+\frac{|n|}{r}\right) \\ 0 & \frac{2(1+\nu)}{h^*}\frac{E^*}{E} & 0 & 0 & 0 & \frac{1}{2}\left(\frac{\partial}{\partial r}-\frac{|n|}{r}\right) \\ -\frac{2(1+\nu)(1-2\nu)}{(1-\nu)h^*}\frac{E^*}{E} & 0 & 0 & \frac{\nu}{1-\nu}\left(\frac{\partial}{\partial r}+\frac{1-|n|}{r}\right) & \frac{\nu}{1-\nu}\left(\frac{\partial}{\partial r}+\frac{1+|n|}{r}\right) & 0 \end{bmatrix} \tag{3.18}$$

und

$$\hat{\boldsymbol{B}} = \begin{bmatrix} \frac{2\nu}{1-\nu} & 0 & 0 & \frac{h^*}{1-\nu}\frac{E}{E^*}\frac{|n|}{r} & \frac{h^*}{1-\nu}\frac{E}{E^*}\left(\frac{\partial}{\partial r}+\frac{1}{r}\right) & 0 \\ 0 & 0 & 0 & \frac{-h^*}{1+\nu}\frac{E}{E^*}\frac{|n|}{r} & \frac{h^*}{1+\nu}\frac{E}{E^*}\left(\frac{\partial}{\partial r}-\frac{1}{r}\right) & 0 \\ 0 & 0 & 0 & \frac{h^*}{1+\nu}\frac{E}{E^*}\left(\frac{\partial}{\partial r}-\frac{1}{r}\right) & \frac{-h^*}{1+\nu}\frac{E}{E^*}\frac{|n|}{r} & 0 \end{bmatrix}. \tag{3.11}$$

Damit ist die Reduktion der partiellen Differentialgleichungen (2.15) in r, φ, z auf solche in r, z vollzogen (Gl. (3.6)).

b) Zweiter Reduktionsschritt: Anwendung der Hankel-Transformation

Um (3.6) auf gewöhnliche Differentialgleichungen in z und um (3.7) auf algebraische Gleichungen zu reduzieren, wenden wir die Hankel-Transformation an. Dies ist aber erst nach Bildung geeigneter Kombinationen der Einzelgleichungen möglich. Man benötigt hierzu die beiden Transformationsmatrizen

$$\boldsymbol{R} = \begin{bmatrix} 1 & & & & & \\ & 1 & 1 & & & \\ & 1 & -1 & & & \\ & & & -1 & 1 & \\ & & & 1 & 1 & \\ & & & & & 2 \end{bmatrix} \quad \text{und} \quad \boldsymbol{R}' = \begin{bmatrix} 1 & 0 & 0 \\ 0 & 1 & 1 \\ 0 & 1 & -1 \end{bmatrix}. \tag{3.12), (3.13}$$

Damit bildet man die neuen Spaltenvektoren gemäß

$$\begin{aligned} \tilde{\boldsymbol{a}} = \boldsymbol{R}\,\hat{\boldsymbol{a}} &= \{\sigma, \tau_I, \tau_{II}, u_{II}, u_I, u\}^T \\ &= \{\hat{\sigma}_{zz}, \hat{\sigma}_{zr} + \hat{\hat{\sigma}}_{z\varphi}, \hat{\sigma}_{zr} - \hat{\hat{\sigma}}_{z\varphi}, \hat{u}_r^* - \hat{\hat{u}}_\varphi^*, \hat{u}_r^* + \hat{\hat{u}}_\varphi^*, -2\,\hat{u}_z^*\}^T \end{aligned} \tag{3.14}$$

und

$$\begin{aligned} \tilde{\boldsymbol{b}} = \boldsymbol{R}'\,\hat{\boldsymbol{b}} &= \{\sigma_I, \sigma_{II}, \sigma_{III}\}^T \\ &= \{(\sigma_{rr} + \sigma_{\varphi\varphi})^\wedge, (\sigma_{rr} - \sigma_{\varphi\varphi})^\wedge + 2\,\hat{\hat{\sigma}}_{r\varphi}, (\sigma_{rr} - \sigma_{\varphi\varphi})^\wedge - 2\,\hat{\hat{\sigma}}_{r\varphi}\}^T . \end{aligned} \tag{3.15}$$

Die Gl. (3.6) und (3.7) gehen dann über in

$$\frac{\partial}{\partial z}\tilde{\boldsymbol{a}} = \tilde{\boldsymbol{A}}\,\tilde{\boldsymbol{a}} \qquad \text{mit} \qquad \tilde{\boldsymbol{A}} = \boldsymbol{R}\,\hat{\boldsymbol{A}}\,\boldsymbol{R}^{-1}, \tag{3.16}$$

$$\tilde{\boldsymbol{b}} = \tilde{\boldsymbol{B}}\,\tilde{\boldsymbol{a}} \qquad \text{mit} \qquad \tilde{\boldsymbol{B}} = \boldsymbol{R}'\,\hat{\boldsymbol{B}}\,\boldsymbol{R}^{-1}, \tag{3.17}$$

ausführlich:

$$\tilde{\boldsymbol{B}} = \begin{bmatrix} \frac{2\nu}{1-\nu} & 0 & 0 & \frac{h^*}{2(1-\nu)}\frac{E}{E^*}\left(\frac{\partial}{\partial r}+\frac{1-|n|}{r}\right) & \frac{h^*}{2(1-\nu)}\frac{E}{E^*}\left(\frac{\partial}{\partial r}+\frac{1+|n|}{r}\right) & 0 \\ 0 & 0 & 0 & 0 & \frac{h^*}{1+\nu}\frac{E}{E^*}\left(\frac{\partial}{\partial r}-\frac{1+|n|}{r}\right) & 0 \\ 0 & 0 & 0 & \frac{h^*}{1+\nu}\frac{E}{E^*}\left(\frac{\partial}{\partial r}+\frac{|n|-1}{r}\right) & 0 & 0 \end{bmatrix}. \tag{3.19}$$

Nach diesen Vorbereitungen führen wir die Hankel-Transformation n-ter Ordnung ein [4]:

$$\overset{n}{f}(\lambda) = \mathcal{H}_n\{f(r)\} = \int_0^\infty r\, f(r)\, I_n(\lambda\, r)\, dr\,. \tag{3.20}$$

Hierin bezeichnet λ den Transformationsparameter und I_n die Besselsche Funktion erster Art und n-ter Ordnung. Die inverse Transformierte ist

$$f(r) = \mathcal{H}_n^{-1}\{\overset{n}{f}(\lambda)\} = \int_0^\infty \lambda\, \overset{n}{f}(\lambda)\, I_n(\lambda\, r)\, d\lambda\,. \tag{3.21}$$

Dabei wird vorausgesetzt, daß $f(r)$ in der Nachbarschaft des Punktes r von beschränkter Schwankung ist und das Integral $\int_0^\infty f(r)\, dr$ absolut konvergiert. An einer Sprungstelle erhält man $\frac{1}{2}[f(r+0) + f(r-0)]$ anstelle von $f(r)$. Für die Hankel-Transformierten der Ableitungen der Funktion $f(r)$ bzw. einiger Kombinationen gelten die im folgenden zusammengestellten

Relationen:

$$\left.\begin{aligned}
&\overset{n}{\overline{\left(\frac{d}{dr}+\frac{1\pm n}{r}\right)f(r)}} = \pm\lambda\,\overset{n\pm1}{\overline{f(\lambda)}}\\
&\overset{n\pm1}{\overline{\left(\frac{d}{dr}\mp\frac{n}{r}\right)f(r)}} = \mp\lambda\,\overset{n}{\overline{f(\lambda)}}\\
&\overset{n\pm2}{\overline{\left(\frac{d}{dr}-\frac{1\pm n}{r}\right)f(r)}} = \mp\lambda\,\overset{n\pm1}{\overline{f(\lambda)}}
\end{aligned}\right\}\quad \text{wenn } r\,f(r)\to 0 \quad \text{für } r\to 0,\ r\to\infty\,; \tag{3.22}$$

$$\left.\begin{aligned}
&\overset{n\pm1}{\overline{\left(\frac{d^2}{dr^2}+\frac{1}{r}\frac{d}{dr}-\frac{(1\pm n)^2}{r^2}\right)f(r)}} = -\lambda^2\,\overset{n\pm1}{\overline{f(\lambda)}}\\
&\overset{n\pm1}{\overline{\left(\frac{d^2}{dr^2}+\frac{1\mp 2n}{r}\frac{d}{dr}-\frac{1-n^2}{r^2}\right)f(r)}} = \lambda^2\,\overset{n\mp1}{\overline{f(\lambda)}}
\end{aligned}\right\}\quad \begin{aligned}&\text{wenn } r\,f(r)\to 0\,,\ r\frac{df}{dr}\to 0\\ &\text{und } I_{n\pm1}\cdot f(r)\to 0\\ &\text{für } r\to 0,\ r\to\infty\,.\end{aligned} \tag{3.23}$$

Zum Beweis gehen wir von der bekannten Beziehung

$$\frac{dI_n(\lambda r)}{d(\lambda r)} = \left\langle\begin{matrix} -\dfrac{n}{\lambda r} I_n + I_{n-1}\\[2mm] \dfrac{n}{\lambda r} I_n - I_{n+1}\end{matrix}\right. \tag{a}$$

aus. Durch partielles Integrieren folgt

$$\int_0^\infty r\frac{df}{dr} I_n(\lambda r)\,dr = \int_0^\infty f(r)\left\langle\begin{matrix}(n-1)\,I_n - \lambda r\,I_{n-1}\\ -(n+1)\,I_n + \lambda r\,I_{n+1}\end{matrix}\right\rangle dr\,, \tag{b}$$

falls $r\,f(r)\to 0$ für $r\to 0$ und $r\to\infty$ sowie

$$\int_0^\infty \frac{df}{dr} I_n(\lambda r)\,dr = \lambda\int_0^\infty f(r)\left\langle\begin{matrix}\dfrac{n}{\lambda r} I_n - I_{n-1}\\[2mm] -\dfrac{n}{\lambda r} I_n + I_{n+1}\end{matrix}\right\rangle dr\,, \tag{c}$$

falls $I_n(\lambda r)\,f(r)\to 0$ für $r\to 0$ und $r\to\infty$. Damit ergibt sich

$$\int_0^\infty r\frac{d^2f}{dr^2} I_n(\lambda r)\,dr = \lambda\int_0^\infty f(r)\left\langle\begin{matrix}\left[\dfrac{n(n-1)}{\lambda r}-\lambda r\right] I_n + I_{n-1}\\[2mm] \left[\dfrac{n(n+1)}{\lambda r}-\lambda r\right] I_n - I_{n+1}\end{matrix}\right\rangle dr\,, \tag{d}$$

falls $r\,f(r)\to 0$, $r\dfrac{df}{dr}\to 0$ und $I_n(\lambda r)\,f(r)\to 0$ für $r\to 0$ und $r\to\infty$.

Die Beziehungen (3.22) erhält man unmittelbar aus (b), und die Beziehungen (3.23) mit Hilfe von (c), (d) und (a).

Die Anwendung der Operationen

$$\boldsymbol{H} = \operatorname{diag}\left(\mathcal{H}_{|n|}, \mathcal{H}_{|n|+1}, \mathcal{H}_{|n|-1}, \mathcal{H}_{|n|-1}, \mathcal{H}_{|n|+1}, \mathcal{H}_{|n|}\right) \tag{3.24}$$

und

$$\boldsymbol{H}' = \operatorname{diag}\left(\mathcal{H}_{|n|}, \mathcal{H}_{|n|+2}, \mathcal{H}_{|n|-2}\right) \tag{3.25}$$

auf (3.16) bzw. (3.17) ergibt zunächst

$$\frac{d}{dz}\boldsymbol{H}\,\tilde{\boldsymbol{a}} = \boldsymbol{H}\,\tilde{\boldsymbol{A}}\,\tilde{\boldsymbol{a}}\,,\qquad \boldsymbol{H}'\,\tilde{\boldsymbol{b}} = \boldsymbol{H}'\,\tilde{\boldsymbol{B}}\,\tilde{\boldsymbol{a}}$$

und bei Beachtung von (3.22) und (3.23) schließlich

$$\frac{d}{dz}\bar{\boldsymbol{a}} = \bar{\boldsymbol{A}}\,\bar{\boldsymbol{a}}\,,\qquad \bar{\boldsymbol{b}} = \bar{\boldsymbol{B}}\,\bar{\boldsymbol{a}}\,. \tag{3.26), (3.27}$$

Die hierin vorkommenden Spaltenvektoren $\bar{\boldsymbol{a}}$, $\bar{\boldsymbol{b}}$ und die Matrizen $\bar{\boldsymbol{A}}$, $\bar{\boldsymbol{B}}$ haben folgende Bedeutung:

$$\bar{\boldsymbol{a}} = \boldsymbol{H}\,\tilde{\boldsymbol{a}} = \left\{\overset{|n|}{\overline{\sigma}},\ \overset{|n|+1}{\overline{\tau_I}},\ \overset{|n|-1}{\overline{\tau_{II}}},\ \overset{|n|-1}{\overline{u_{II}}},\ \overset{|n|+1}{\overline{u_I}},\ \overset{|n|}{\overline{u}}\right\}^T, \tag{3.28}$$

$$\bar{\boldsymbol{b}} = \boldsymbol{H}'\,\tilde{\boldsymbol{b}} = \left\{\overset{|n|}{\overline{\sigma_I}},\ \overset{|n|+2}{\overline{\sigma_{II}}},\ \overset{|n|-2}{\overline{\sigma_{III}}}\right\}^T, \tag{3.29}$$

$$\overline{A} = \begin{bmatrix} 0 & -\frac{\lambda^*}{2\,h^*} & \frac{\lambda^*}{2\,h^*} & 0 & 0 & 0 \\ \frac{\nu}{1-\nu}\frac{\lambda^*}{h^*} & 0 & 0 & -\frac{1}{4\,(1-\nu)}\frac{E}{E^*}\frac{\lambda^{*2}}{h^*} & \frac{3-\nu}{4\,(1-\nu^2)}\frac{E}{E^*}\frac{\lambda^{*2}}{h^*} & \\ \frac{-\nu}{1-\nu}\frac{\lambda^*}{h^*} & 0 & 0 & \frac{3-\nu}{4\,(1-\nu^2)}\frac{E}{E^*}\frac{\lambda^{*2}}{h^*} & & \\ 0 & 0 & \frac{2\,(1+\nu)}{h^*}\frac{E^*}{E} & & & \\ 0 & \frac{2\,(1+\nu)}{h^*}\frac{E^*}{E} & & \text{symmetrisch} & & \\ -\frac{2(1+\nu)(1-2\nu)}{(1-\nu)\,h^*}\frac{E^*}{E} & & & & & \end{bmatrix}, \tag{3.30}$$

$$\overline{B} = \begin{bmatrix} \frac{2\,\nu}{1-\nu} & 0 & 0 & -\frac{\lambda^*}{2\,(1-\nu)}\frac{E}{E^*} & \frac{\lambda^*}{2\,(1-\nu)}\frac{E}{E^*} & 0 \\ 0 & 0 & 0 & 0 & -\frac{\lambda^*}{1+\nu}\frac{E}{E^*} & 0 \\ 0 & 0 & 0 & \frac{\lambda^*}{1+\nu}\frac{E}{E^*} & 0 & 0 \end{bmatrix}. \tag{3.31}$$

Ferner ist

$$\lambda^* = \lambda\, h^* \,. \tag{3.32}$$

Gl. (3.26) ist eine gewöhnliche Matrizendifferentialgleichung 1. Ordnung für den „Zustandsvektor" $\overline{a}$, während (3.27) die (algebraische) Transformation von $\overline{a}$ auf $\overline{b}$ leistet.

3.2. Übertragungsgleichung und Übertragungsmatrix[1])

Die Integration von (3.26) liefert mit dem „Anfangsvektor" $\overline{a}(0)$ an der Stelle $z^* = \frac{z}{h^*} = 0$ den Zustandsvektor $\overline{a}(z^*)$ an der beliebigen Stelle z^* der betrachteten Schicht:

$$\overline{a}(z^*) = T(z^*)\,\overline{a}(0)\,. \tag{3.33}$$

Die quadratische 6×6 Matrix T ist dabei gegeben durch die unendliche Reihe

$$T = e^{\overline{A}\,z^*\,h^*} = I + \overline{A}\,z^*\,h^* + \frac{1}{2!}(\overline{A}\,z^*\,h^*)^2 + \cdots, \tag{3.34}$$

die man wegen des CAYLEY-HAMILTONschen Theorems durch ein Polynom 5. Grades ersetzen kann:

$$T = c_0\,I + c_1\,\overline{A}\,z^*\,h^* + c_2\,(\overline{A}\,z^*\,h^*)^2 + \cdots + c_5(\overline{A}\,z^*\,h^*)^5\,. \tag{3.35}$$

Die Koeffizienten c_0 bis c_5 errechnen sich auf Grund der Eigenwerte ϱ_i der Matrix $\overline{A}\,z^*\,h^*$

$$\varrho_{1,2,3} = \lambda^*\,z^*\,, \qquad \varrho_{4,5,6} = -\lambda^*\,z^*\,.$$

Das Ergebnis lautet mit

$$s(z^*) = \sinh(\lambda^*\,z^*)\,, \qquad c(z^*) = \cosh(\lambda^*\,z^*)$$

$$\begin{aligned} c_0 &= c(z^*) - \frac{5}{8}\lambda^*\,z^*\,s(z^*) + \frac{1}{8}(\lambda^*\,z^*)^2\,c(z^*)\,, \\ c_1 &= \frac{1}{8\lambda^*\,z^*}[15\,s(z^*) - 7\,\lambda^*\,z^*\,c(z^*) + (\lambda^*\,z^*)^2\,s(z^*)]\,, \\ c_2 &= \frac{1}{4\,\lambda^*\,z^*}[3\,s(z^*) - \lambda^*\,z^*\,c(z^*)]\,, \\ c_3 &= -\frac{1}{4(\lambda^*\,z^*)^3}[5\,\langle s(z^*) - \lambda^*\,z^*\,c(z^*)\rangle + (\lambda^*\,z^*)^2\,s(z^*)]\,, \\ c_4 &= -\frac{1}{8(\lambda^*\,z^*)^3}[s(z^*) - \lambda^*\,z^*\,c(z^*)]\,, \\ c_5 &= \frac{1}{8(\lambda^*\,z^*)^5}[3\,\langle s(z^*) - \lambda^*\,z^*\,c(z^*)\rangle + (\lambda^*\,z^*)^2\,s(z^*)]\,. \end{aligned} \tag{3.36}$$

[1]) Zum Verfahren der Übertragungsmatrizen siehe z. B. PESTEL u. LECKIE [7].

$$
\boldsymbol{T}_k(z_k^*) = \frac{c(z_k^*)}{2\,(1-\nu_k)}
\left[\begin{array}{cccccc}
2\,(1-\nu_k) - \lambda^* z_k^*\, t(z_k^*) & -\frac{1}{2}\,[\lambda^* z_k^* + (1-2\,\nu_k)\, t(z_k^*)] & \frac{1}{2}\,[\lambda^* z_k^* + (1-2\,\nu_k)\, t(z_k^*)] & \frac{E_k^*}{2\,(1+\nu_k)}\lambda^{*2} z_k^*\, t(z_k^*) & -\frac{E_k^*\lambda^{*2}}{2\,(1+\nu_k)}\, z_k^*\, t(z_k^*) & \frac{E_k^*\lambda^*}{2\,(1+\nu_k)}\,[\lambda^* z_k^* - t(z_k^*)] \\
\lambda^* z_k^* - (1-2\,\nu_k)\, t(z_k^*) & 2\,(1-\nu_k) + \frac{\lambda^* z_k^*}{2}\, t(z_k^*) & -\frac{\lambda^* z_k^*}{2}\, t(z_k^*) & -\frac{E_k^*\lambda^*}{2\,(1+\nu_k)}\,[\lambda^* z_k^* + \nu_k\, t(z_k^*)] & \frac{E_k^*\lambda^*}{2\,(1+\nu_k)}\,[\lambda^* z_k^* + (2-\nu_k)\, t(z_k^*)] & \\
-\lambda^* z_k^* + (1-2\,\nu_k)\, t(z_k^*) & -\frac{\lambda^* z_k^*}{2}\, t(z_k^*) & 2\,(1-\nu_k) + \frac{\lambda^* z_k^*}{2}\, t(z_k^*) & \frac{E_k^*\lambda^*}{2\,(1+\nu_k)}\,[\lambda^* z_k^* + (2-\nu_k)\, t(z_k^*)] & & \\
-\frac{1+\nu_k}{E_k^*}\, z_k^*\, t(z_k^*) & -\frac{1+\nu_k}{2\,E_k^*}\left[z_k^* - \frac{t(z_k^*)}{\lambda^*}\right] & \frac{1+\nu_k}{2\,E_k^*}\left[z_k^* + (7-8\,\nu_k)\,\frac{t(z_k^*)}{\lambda^*}\right] & & & \\
\frac{1+\nu_k}{E_k^*}\, z_k^*\, t(z_k^*) & \frac{1+\nu_k}{2\,E_k^*}\left[z_k^* + (7-8\,\nu_k)\,\frac{t(z_k^*)}{\lambda^*}\right] & & \text{symmetrisch} & & \\
\frac{2\,(1+\nu_k)}{E_k^*}\left[z_k^* - \frac{3-4\,\nu_k}{\lambda^*}\, t(z_k^*)\right] & & & & &
\end{array}\right]
\tag{3.37}
$$

$$
{}^{(0)}\boldsymbol{T}_k(z_k^*) =
\left[\begin{array}{cccccc}
1 & -\frac{1}{2}\lambda^* z_k^* & \frac{1}{2}\lambda^* z_k^* & \frac{E_k^*}{4\,(1-\nu_k^2)}\lambda^{*3} z_k^{*2} & -\frac{E_k^*\lambda^{*3}}{4\,(1-\nu_k^2)}\, z_k^{*2} & \frac{E_k^*\lambda^{*4}}{12\,(1-\nu_k^2)}\, z_k^{*3} \\
\frac{\nu_k}{1-\nu_k}\lambda^* z_k^* & 1 & -\frac{\lambda^{*2}}{4\,(1-\nu_k)}\, z_k^{*2} & -\frac{E_k^*\lambda^{*2}}{4\,(1-\nu_k)}\, z_k^* & \frac{3-\nu_k}{4\,(1-\nu_k^2)}\, E_k^*\lambda^{*2} z_k^* & \\
\frac{-\nu_k}{1-\nu_k}\lambda^* z_k^* & -\frac{\lambda^{*2}}{4\,(1-\nu_k)}\, z_k^{*2} & 1 & \frac{3-\nu_k}{4\,(1-\nu_k^2)}\, E_k^*\lambda^{*2} z_k^* & & \\
-\frac{1+\nu_k}{2\,E_k^*\,(1-\nu_k)}\lambda^* z_k^{*2} & -\frac{(1+\nu_k)\,\lambda^{*2}}{12\,(1-\nu_k)\,E_k^*}\, z_k^{*3} & \frac{2\,(1+\nu_k)}{E_k^*}\, z_k^* & & & \\
\frac{1+\nu_k}{2\,E_k^*\,(1-\nu_k)}\lambda^* z_k^{*2} & \frac{2\,(1+\nu_k)}{E_k^*}\, z_k^* & & \text{symmetrisch} & & \\
-\frac{2\,(1+\nu_k)\,(1-2\,\nu_k)}{(1-\nu_k)\,E_k^*}\, z_k^* & & & & &
\end{array}\right].
\tag{3.42}
$$

Einsetzen von (3.36) in (3.35) und Zusammenfassen liefert die Matrix (3.37), in der der Index k auf die Schicht k hinweisen soll und die Abkürzungen

$$z_k^* = \frac{z_k}{h^*}, \qquad E_k^* = \frac{E_k}{E^*},$$
$$c(z_k^*) = \cosh(\lambda^* z_k^*), \qquad t(z_k^*) = \tanh(\lambda^* z_k^*) \tag{3.38}$$

verwendet wurden.

Diese Matrix hat die bekannten Eigenschaften einer Übertragungsmatrix: Symmetrie zur Nebendiagonale, wenn die Elemente des Zustandsvektors in einer geeigneten Reihenfolge gewählt werden, Determinante 1, gleiche Elemente mit schachbrettartiger Vorzeichenvertauschung für die inverse Matrix; ferner ist (innerhalb einer Schicht)

$$\boldsymbol{T}(z_1^* + z_2^*) = \boldsymbol{T}(z_2^*)\,\boldsymbol{T}(z_1^*)\,.$$

Die Matrix $\boldsymbol{T}_k(z_k^*)$ erlaubt gemäß (3.33) die Berechnung von $\overline{\boldsymbol{a}}_k(z_k^*)$, wenn man $\overline{\boldsymbol{a}}_k(0)$ kennt:

$$\overline{\boldsymbol{a}}_k(z_k^*) = \boldsymbol{T}_k(z_k^*)\,\overline{\boldsymbol{a}}_k(0)\,. \tag{3.39}$$

Ferner erhält man gemäß (3.27) und (3.39)

$$\overline{\boldsymbol{b}}_k(z_k^*) = \overline{\boldsymbol{B}}_k\,\overline{\boldsymbol{a}}_k(z_k^*) = \overline{\boldsymbol{B}}_k\,\boldsymbol{T}_k(z_k^*)\,\overline{\boldsymbol{a}}_k(0) \tag{3.40}$$

worin

$$\overline{\boldsymbol{B}}_k = \overline{\boldsymbol{B}} \quad \text{mit} \quad \nu = \nu_k\,, \quad \frac{E}{E^*} = \frac{E_k}{E^*} = E_k^*\,. \tag{3.41}$$

Eine *dünne Schicht* sei dadurch charakterisiert, daß für sie $(\lambda^* z_k^*)^2 \ll 1$ gilt, d. h. die „Wellenlänge" der Belastung sei groß im Vergleich zur betrachteten Koordinate $0 \leqq z_k \leqq h_k$. In diesem Fall liefert die TAYLOR-Entwicklung der Elemente von $\boldsymbol{T}_k(z_k^*)$, wenn man jeweils nur das erste Glied beibehält, die in (3.42) wiedergegebene Matrix ${}^{(0)}\boldsymbol{T}_k(z_k^*)$. Man kann diese übrigens auch aus (3.34) unter Umgehung der Eigenwertbestimmung herleiten.

Setzt man in (3.37) bzw. (3.42) $z_k = h_k$, d. h. $z_k^* = h_k^* = h_k/h^*$, so erhält man die Übertragungsmatrix der Schicht k:

$$\boldsymbol{T}_k = \boldsymbol{T}_k(h_k^*) \quad \text{bzw.} \quad {}^{(0)}\boldsymbol{T}_k = {}^{(0)}\boldsymbol{T}_k(h_k^*)\,. \tag{3.43}$$

Es läßt sich nachweisen, daß die Übertragungsmatrix der dünnen Schicht die dünne Platte mit Berücksichtigung der Schubdeformation sowie die dünne Scheibe (Membranzustand) beschreibt.

3.3. Nachgiebigkeitsmatrix

Interessiert man sich nur für die Spannungen eines Schichtsystems, so ist es zweckmäßig, die Verschiebungsgrößen zu eliminieren. Man hat hierzu die Übertragungsgleichung

$$\overline{\boldsymbol{a}}_k(h_k^*) = \boldsymbol{T}_k\,\overline{\boldsymbol{a}}_k(0) \tag{3.44}$$

folgendermaßen umzuformen:

$$\begin{bmatrix} \overline{\boldsymbol{u}}(0) \\ \overline{\boldsymbol{u}}(h) \end{bmatrix}_k = \begin{bmatrix} \boldsymbol{N}_{11} & \boldsymbol{N}_{12} \\ \boldsymbol{N}_{21} & \boldsymbol{N}_{22} \end{bmatrix}_k \begin{bmatrix} -\overline{\boldsymbol{\sigma}}(0) \\ \overline{\boldsymbol{\sigma}}(h) \end{bmatrix}_k . \tag{3.45}$$

Ausführlich:

$$\begin{bmatrix} \overset{|n|+1}{\overline{u}}_{\mathrm{I}}(0) \\ \overset{|n|-1}{\overline{u}}_{\mathrm{II}}(0) \\ \overset{|n|}{\overline{u}}(0) \\ \overset{|n|+1}{\overline{u}}_{\mathrm{I}}(h) \\ \overset{|n|-1}{\overline{u}}_{\mathrm{II}}(h) \\ \overset{|n|}{\overline{u}}(h) \end{bmatrix}_k = q \cdot \begin{bmatrix} a & b & c & d & e & f \\ & a & -c & e & d & -f \\ & & g & -f & f & h \\ & & & a & b & -c \\ & \text{symmetrisch} & & & a & c \\ & & & & & g \end{bmatrix}_k \begin{bmatrix} -\overset{|n|+1}{\overline{\tau}}_{\mathrm{I}}(0) \\ -\overset{|n|-1}{\overline{\tau}}_{\mathrm{II}}(0) \\ \overset{|n|}{\overline{\sigma}}(0) \\ \overset{|n|+1}{\overline{\tau}}_{\mathrm{I}}(h) \\ \overset{|n|-1}{\overline{\tau}}_{\mathrm{II}}(h) \\ -\overset{|n|}{\overline{\sigma}}(h) \end{bmatrix}_k \tag{3.46}$$

mit

	Normalschicht	dünne Schicht
a	$(1-\nu_k)\, t_k \left[t_k - \frac{\lambda^* h_k^*}{c_k^2}\right] + t_k^2 - \left(\frac{\lambda^* h_k^*}{c_k}\right)^2$	$(3-2\nu_k)(\lambda^* h_k^*)^2$
b	$(1-\nu_k)\, t_k \frac{\lambda^* h_k^*}{c_k^2} + \nu_k t_k^2 - \left(\frac{\lambda^* h_k^*}{c_k}\right)^2$	$-(1-2\nu_k)(\lambda^* h_k^*)^2$
c	$t_k \left[(1-2\nu_k)\, t_k^2 + \left(\frac{\lambda^* h_k^*}{c_k}\right)^2\right]$	$6(1-\nu_k)\,\lambda^* h_k^*$
d	$(1-\nu_k) \frac{t_k}{c_k} [t_k - \lambda^* h_k^*] + \frac{1}{c_k}\left[t_k^2 - \left(\frac{\lambda^* h_k^*}{c_k}\right)^2\right]$	$\nu_k (\lambda^* h_k^*)^2$
e	$(1-\nu_k)\, t_k \frac{\lambda^* h_k^*}{c_k} + \frac{1}{c_k}\left[\nu_k t_k^2 - \left(\frac{\lambda^* h_k^*}{c_k}\right)^2\right]$	$(2-\nu_k)(\lambda^* h_k^*)^2$
f	$2(1-\nu_k)\, t_k^2 \frac{\lambda^* h_k^*}{c_k}$	$6(1-\nu_k)\,\lambda^* h_k^*$
g	$4(1-\nu_k)\, t_k \left[t_k + \frac{\lambda^* h_k^*}{c_k^2}\right]$	$24(1-\nu_k)$
h	$4(1-\nu_k) \frac{t_k}{c_k} [t_k + \lambda^* h_k^*]$	$24(1-\nu_k)$
q	$\dfrac{(1+\nu_k)}{E_k^* \lambda^* t_k \left[t_k^2 - \left(\frac{\lambda^* h_k^*}{c_k}\right)^2\right]}$	$\dfrac{1+\nu_k}{E_k^* \lambda^* (\lambda^* h_k^*)^3}$.

(3.47)

$$t_k = \tanh(\lambda^* h_k^*)\,, \qquad c_k = \cosh(\lambda^* h_k^*)\,.$$

Die Vorzeichenfestsetzung bei $\bar{\sigma}(0)$ in (3.45) ist darin begründet, daß die positiven Richtungen der Kraftgrößen mit den positiven Richtungen der Verschiebungsgrößen übereinstimmen sollen. Dann ergibt sich auch die durch den MAWXELLschen Reziprozitätssatz ausgedrückte Symmetrie der Nachgiebigkeitsmatrix. Diese hat nur 8 voneinander verschiedene Elemente, im Gegensatz zur Übertragungsmatrix mit 13 voneinander verschiedenen Elementen.

4. Das Mehrschichtsystem im Falle vollständiger Haftung aller Schichten

4.1. Das Übertragungsverfahren

Im Falle vollständiger Haftung müssen die Spannungskomponenten σ_{rr}, $\sigma_{z\varphi}$, σ_{zz} und die Verschiebungskomponenten u_r, u_φ, u_z der jeweils angrenzenden Schichten stetig ineinander übergehen. Dies gilt selbstverständlich auch für die Transformierten dieser Größen bzw. ihrer Kombinationen, so daß wir die Übergangsbedingungen in der Form

$$\bar{\boldsymbol{a}}_{k+1}(0) = \bar{\boldsymbol{a}}_k(h_k) = \bar{\boldsymbol{a}}_k \qquad (k = 1, 2, \ldots, N-1) \tag{4.1}$$

schreiben können; $\bar{\boldsymbol{a}}_k$ bedeutet also den (eindeutigen) Zustandsvektor an der Stelle k.

Die Übertragungsgleichung (3.44) unter Berücksichtigung von (4.1) lautet für die Stellen $k-1$ und k

$$\bar{\boldsymbol{a}}_k = \boldsymbol{T}_k\, \bar{\boldsymbol{a}}_{k-1} \qquad (k = 1, 2, \ldots, N)\,. \tag{4.2}$$

Ihre successive Anwendung

$$\bar{\boldsymbol{a}}_1 = \boldsymbol{T}_1\, \bar{\boldsymbol{a}}_0\,, \qquad \bar{\boldsymbol{a}}_2 = \boldsymbol{T}_2\, \bar{\boldsymbol{a}}_1\,, \qquad \bar{\boldsymbol{a}}_N = \boldsymbol{T}_N\, \bar{\boldsymbol{a}}_{N-1} \tag{4.3}$$

erlaubt die Elimination der Zwischengrößen und ergibt

$$\bar{\boldsymbol{a}}_N = \boldsymbol{T}_N \cdots \boldsymbol{T}_1\, \bar{\boldsymbol{a}}_0 = \boldsymbol{T}\, \bar{\boldsymbol{a}}_0\,, \tag{4.4}$$

also eine Verknüpfung der Randgrößen. Ausführlich:

$$\begin{bmatrix} \overset{|n|}{\sigma} \\ \overset{|n|+1}{\tau_I} \\ \overset{|n|-1}{\tau_{II}} \\ \overset{|n|-1}{u_{II}} \\ \overset{|n|+1}{u_I} \\ \overset{|n|}{u} \end{bmatrix}_N = \begin{bmatrix} t_{11} & t_{12} & t_{13} & t_{14} & t_{15} & t_{16} \\ t_{21} & t_{22} & t_{23} & t_{24} & & \cdot \\ t_{31} & t_{32} & & & & \cdot \\ \cdot & & & & & \cdot \\ \cdot & & & & & \cdot \\ \cdot & \cdot & & & & \cdot \end{bmatrix} \begin{bmatrix} \overset{|n|}{\sigma} \\ \overset{|n|+1}{\tau_I} \\ \overset{|n|-1}{\tau_{II}} \\ \overset{|n|-1}{u_{II}} \\ \overset{|n|+1}{u_I} \\ \overset{|n|}{u} \end{bmatrix}_0 . \tag{4.5}$$

Drei Komponenten von $\bar{a}_0$ und drei Komponenten von $\bar{a}_N$ sind auf Grund der Randbedingungen als bekannt anzusehen. Genauer: Sie sind über (2.13), (3.8), (3.14) und (3.28) aus den bekannten Randgrößen zu berechnen:

$$\bar{a} = \boldsymbol{H}\,\tilde{a} = \boldsymbol{H}\,\boldsymbol{R}\,\hat{a} = \boldsymbol{H}\,\boldsymbol{R}\,\boldsymbol{F}\,\boldsymbol{a} . \tag{4.6}$$

Aus den sechs Gleichungen (4.5) folgen die restlichen Komponenten. Mit $\bar{a}_0$ stehen aber wegen (4.3) sämtliche $\bar{a}_{k-1} = \bar{a}_k(0)$ fest. Gl. (3.39) und (3.40) liefern dann $\bar{a}_k(z_k^*)$ und $\bar{b}_k(z_k^*)$. Die Rücktransformationen führen schließlich zum vollständigen Spannungs- und Verschiebungszustand:

$$\boldsymbol{a}_k(r, \varphi, z_k) = \boldsymbol{F}^{-1}\{\boldsymbol{R}^{-1}\,\boldsymbol{H}^{-1}\langle \boldsymbol{T}_k(z_k^*)\,\bar{a}_k(0)\rangle\} , \tag{4.7}$$

$$\boldsymbol{b}_k(r, \varphi, z_k) = \boldsymbol{F}'^{-1}\{\boldsymbol{R}'^{-1}\,\boldsymbol{H}'^{-1}\langle \bar{\boldsymbol{B}}_k\,\boldsymbol{T}_k(z_k^*)\,\bar{a}_k(0)\rangle\} . \tag{4.8}$$

Falls die letzte Schicht N sehr dick ist (*Halbraum*, $h_N \to \infty$), besteht eine Kopplung zwischen den 6 Anfangsgrößen in $\bar{a}_{N-1}$:

$$\boldsymbol{S}_N\,\bar{a}_{N-1} = 0 . \tag{4.9}$$

Hierin bedeutet

$$\boldsymbol{S}_N = \begin{bmatrix} 4(1-\nu_N) & -(1-2\nu_N) & 1-2\nu_N & 0 & 0 & -\dfrac{E_N^*}{1+\nu_N}\lambda^* \\ -(1-2\nu_N) & 2-\nu_N & \nu_N & 0 & \dfrac{E_N^*}{1+\nu_N}\lambda^* & 0 \\ 1-2\nu_N & \nu_N & 2-\nu_N & \dfrac{E_N^*}{1+\nu_N}\lambda^* & 0 & 0 \end{bmatrix} . \tag{4.10}$$

Man gewinnt die Beziehung (4.9) dadurch, daß man die Übertragungsgleichung (4.2) für die Schicht $k = N$ nach den Verschiebungsgrößen am Halbraumrand $N - 1$ auflöst und dann zur Grenze $h_N \to \infty$ übergeht, bei gleichzeitiger Forderung nach einem beschränkten (oder verschwindenden) $\bar{a}_N$. Die Beziehung (4.9) tritt also beim Halbraum an die Stelle der Beziehung $\bar{a}_N = \boldsymbol{T}_N\,\bar{a}_{N-1}$. Bei einem Schichtsystem mit dem Halbraum als letzter Schicht ist demnach die Übertragungsgleichung (4.4) zu ersetzen durch

$$\boldsymbol{S}_N\,\boldsymbol{T}_{N-1} \cdots \boldsymbol{T}_1\,\bar{a}_0 = 0 . \tag{4.11}$$

4.2. Numerische Gesichtspunkte

Die Übertragungsmatrix $\boldsymbol{T}_k = T_k(h_k^*)$ einer Einzelschicht nimmt für sehr große Werte von $\lambda^* h_k^*$ den Rang 1 an; dies gilt ebenso für die Gesamtübertragungsmatrix $\boldsymbol{T}$ eines Schichtsystems. In diesem Fall verschwinden alle Unterdeterminanten von $\boldsymbol{T}_k$ bzw. $\boldsymbol{T}$, und die numerische Auflösung von (4.4) nach den 3 nicht vorgegebenen Elementen von $\bar{a}_0$ muß unbrauchbar werden. Wie ist dieser Effekt zu erklären und wie ist Abhilfe zu schaffen? Zur mechanischen Erklärung dient das de St. Venantsche Prinzip: $\lambda^* h_k^* \gg 1$ bedeutet nämlich, daß die „Wellenlänge" der Belastung in r-Richtung im Vergleich zur Schichtdicke h_k sehr klein ist. Der Einfluß einer an einem Schichtrand wirkenden derartigen „Quasi-Gleichgewichtsgruppe" auf den relativ weit entfernten zweiten Rand ist wegen des Abklingeffektes (de St. Venant) praktisch gleich Null. Folglich müssen zwischen den Zuständen an beiden Rändern numerisch auszuwertende Beziehungen versagen.

Zur mathematischen Erklärung[2]) greifen wir auf die Differentialgleichung (3.26) zurück und integrieren sie durch den Lösungsansatz

$$\bar{a} = \boldsymbol{q}\,e^{\mu z} .$$

Einsetzen führt auf das Eigenwertproblem der Matrix $\bar{\boldsymbol{A}}$. Wenn ein Eigenwert einen positiven Realteil hat — was hier der Fall ist — dann setzt sich für $\lambda z = \lambda^* z^* \gg 1$ der zugehörige Eigenvektor durch, d. h. $\bar{a}(z)$ wird unabhängig vom Anfangsvektor $\bar{a}(0)$.

Hieraus folgt mit Bezug auf die Übertragungsgleichung (3.39), daß alle Zeilen der Übertragungsmatrix $\boldsymbol{T}$ voneinander linear abhängen, d. h. $\boldsymbol{T}$ hat den Rang 1.

---

[2]) Siehe hierzu K. Marguerre und R. Uhrig [8].

Es gibt folgende Möglichkeiten, um Abhilfe zu schaffen:

a) Man entwickelt eine geschlossene Lösung für die Zustandsvektoren, wobei sog. Delta-Matrizen auftreten. Dieser Weg ist für ein Problem 4. Ordnung noch gangbar — ein Beispiel findet sich in [3] — jedoch für Probleme höherer Ordnung wegen des außerordentlich großen Rechenumfanges inpraktikabel.

b) Man löst das bandförmige Gleichungssystem (4.3) simultan nach allen Unbekannten mit Hilfe des GAUSSschen Algorithmus auf, wobei man, falls nötig, die Einzelschichten noch unterteilen kann, und vermeidet damit die Verwendung von (4.4).

c) Man arbeitet mit den Nachgiebigkeitsmatrizen anstelle der Übertragungsmatrizen.

Die Durchrechnung von Zwei- und Dreischichtsystemen zeigt, daß man als obere Integrationsgrenze der bei der HANKELschen Rücktransformation anfallenden und numerisch auszuwertenden Integrale im allgemeinen $\lambda^* = \lambda\, h_k = 15$ mit h_k als Dicke der dünnsten Schicht nehmen darf.

4.3. Das Verfahren der Nachgiebigkeitsmatrizen

Die Übergangsbedingungen, die den Gl. (4.1) entsprechen, schreiben wir hier mit den in (3.45) eingeführten Bezeichnungen in folgender Form:

$$\left.\begin{aligned}\bar{\boldsymbol{\sigma}}_{k+1}(0) &= \bar{\boldsymbol{\sigma}}_k(h_k) = \bar{\boldsymbol{\sigma}}_k\\ \bar{\boldsymbol{u}}_{k+1}(0) &= \bar{\boldsymbol{u}}_k(h_k) = \bar{\boldsymbol{u}}_k\end{aligned}\right\}\quad k = 1, 2, \ldots, N-1 . \tag{4.12}$$

Die erste Beziehung beinhaltet die statischen Übergangsbedingungen und die zweite Beziehung die geometrischen Übergangsbedingungen. Die zweite liefert in Verbindung mit der ersten und mit (3.45)

$$\boldsymbol{N}_{21_k}\,\bar{\boldsymbol{\sigma}}_{k-1} - (\boldsymbol{N}_{11_{k+1}} + \boldsymbol{N}_{22_k})\,\bar{\boldsymbol{\sigma}}_k + \boldsymbol{N}_{12_{k+1}}\,\bar{\boldsymbol{\sigma}}_{k+1} = 0\,, \qquad k = 1, 2, \ldots, N-1 . \tag{4.13}$$

Aus diesem bandförmigen Gleichungssystem lassen sich sämtliche unbekannten Spannungsgrößen $\bar{\boldsymbol{\sigma}}_k$ an den Übergangsstellen errechnen, sofern die Randspannungsgrößen bekannt sind (letztere müssen natürlich insgesamt miteinander im Gleichgewicht stehen). Sind Randverschiebungsgrößen vorgegeben, so hat man diese gemäß (3.45) durch die entsprechenden Randspannungsgrößen auszudrücken. Das Verfahren der Nachgiebigkeitsmatrizen läßt sich auch dann ohne Schwierigkeiten anwenden, wenn eine der Schichten sehr dick ist (Halbraum).

Nach Ermittlung der Spannungsgrößen an den Übergangsstellen folgen die Verschiebungsgrößen aus (3.45). Der vollständige Spannungs- und Verschiebungszustand einer Schicht k errechnet sich schließlich über (3.39) und (3.40).

5. Das glatte Mehrschichtsystem (verschwindende Haftung)

In diesem Fall bestehen für jede Schicht k die Bedingungen

$$\sigma_{zr_{k-1}} = \sigma_{zr_k} = \sigma_{z\varphi_{k-1}} = \sigma_{z\varphi_k} = 0\,, \qquad k = 1, 2, \ldots, N\,, \tag{5.1}$$

oder — ausgedrückt in den transformierten Variablen $\overset{|n|+1}{\bar{\tau}_{\mathrm{I}}}$ und $\overset{|n|-1}{\bar{\tau}_{\mathrm{II}}}$ gemäß (3.28)

$$\overset{|n|+1}{\bar{\tau}}_{\mathrm{I}_{k-1}} = \overset{|n|-1}{\bar{\tau}}_{\mathrm{II}_{k-1}} = \overset{|n|+1}{\bar{\tau}}_{\mathrm{I}_k} = \overset{|n|-1}{\bar{\tau}}_{\mathrm{II}_k} = 0\,, \qquad k = 1, 2, \ldots, N . \tag{5.2}$$

Wir führen nun den „reduzierten Zustandsvektor" $\bar{\boldsymbol{a}}^*$ gemäß

$$\bar{\boldsymbol{a}}^* = \left\{\overset{|n|}{\bar{\sigma}}, -\frac{1}{2}\overset{|n|}{\bar{u}}\right\} = \boldsymbol{D}\,\bar{\boldsymbol{a}} \tag{5.3}$$

ein, wobei

$$\boldsymbol{D} = \begin{bmatrix} 1 & 0 & 0 & 0 & 0 & 0\\ 0 & 0 & 0 & 0 & 0 & -\dfrac{1}{2}\end{bmatrix} \tag{5.4}$$

bedeutet. Aus der Übertragungsgleichung (3.44) folgt wegen (5.2) eine Kopplung zwischen $\overset{|n|}{\bar{\sigma}}_k(0)$, $\overset{|n|}{\bar{u}}_k(0)$, $\overset{|n|+1}{\bar{u}}_{\mathrm{I}_k}(0)$, $\overset{|n|-1}{\bar{u}}_{\mathrm{II}_k}(0)$:

$$\bar{\boldsymbol{a}}_k(0) = \boldsymbol{C}_k\,\bar{\boldsymbol{a}}_k^*(0)\,. \tag{5.5}$$

Die Matrix $\boldsymbol{C}_k$ ist dabei gegeben durch

$$\boldsymbol{C}_k = \begin{bmatrix} 1 & 0\\ 0 & 0\\ 0 & 0\\ \dfrac{1+\nu_k}{E_k^*\,\lambda^*}\,\dfrac{\lambda^*\,h_k^* - (1-2\,\nu_k)\,t_k}{\lambda^*\,h_k^* + t_k} & \dfrac{\lambda^*\,h_k^*\,t_k}{\lambda^*\,h_k^* + t_k}\\ -\dfrac{1+\nu_k}{E_k^*\,\lambda^*}\,\dfrac{\lambda^*\,h_k^* - (1-2\,\nu_k)\,t_k}{\lambda^*\,h_k^* + t_k} & -\dfrac{\lambda^*\,h_k^*\,t_k}{\lambda^*\,h_k^* + t_k}\\ 0 & -2 \end{bmatrix}. \tag{5.6}$$

Gleichung (5.5) leistet mithin die Umrechnung vom reduzierten zum vollständigen Zustandsvektor am „Anfang" ($z_k = 0$) der Schicht k.

Aus (5.3), (3.44) und (5.5) erhalten wir die Übertragungsgleichung für den reduzierten Zustandsvektor:

$$\bar{\boldsymbol{a}}_k^*(h_k) = \boldsymbol{T}_k^* \, \bar{\boldsymbol{a}}_k^*(0) \,. \tag{5.7}$$

Hierin bedeutet

$$\boldsymbol{T}_k^* = \boldsymbol{D} \, \boldsymbol{T}_k \, \boldsymbol{C}_k \tag{5.8}$$

die reduzierte Übertragungsmatrix. Die Ausrechnung liefert

$$\boldsymbol{T}_k^* = \frac{c_k}{(\lambda^* h_k^* + t_k)} \begin{bmatrix} \dfrac{\lambda^* h_k^*}{c_k^2} + t_k & \dfrac{\lambda^* E_k^*}{2\,(1-\nu_k^2)} \left[t_k^2 - \left(\dfrac{\lambda^* h_k^*}{c_k} \right)^2 \right] \\ \dfrac{2\,(1-\nu_k^2)}{E_k^* \, \lambda^*} t_k^2 & \dfrac{\lambda^* h_k^*}{c_k^2} + t_k \end{bmatrix} \tag{5.9}$$

und speziell für die dünne Schicht

$$^{(0)}\boldsymbol{T}_k^* = \begin{bmatrix} 1 & \dfrac{(\lambda^* h_k^*)^4}{12\,(1-\nu_k^2)} \dfrac{E_k^*}{h_k^*} \\ \dfrac{(1-\nu_k^2)\, h_k^*}{E_k^*} & 1 \end{bmatrix} . \tag{5.10}$$

Wir haben wiederum die Eigenschaften: Symmetrie zur Nebendiagonale, Determinante 1, bis auf schachbrettartige Vorzeichenänderung gleiche Elemente in der inversen Matrix. Allerdings gilt jetzt für zwei Schichten aus gleichem Material wegen der Diskontinuität der Verschiebungskomponenten u_r und u_φ

$$\boldsymbol{T}^* (h_1^* + h_2^*) \neq \boldsymbol{T}^*(h_2^*) \, \boldsymbol{T}^*(h_1^*) \,.$$

Die Übergangsbedingungen beziehen sich nunmehr auf die Stetigkeit der Normalspannung σ_{zz} und der Verschiebung u_z beim Übergang von einer Schicht zur Nachbarschicht; mit anderen Worten: Der reduzierte Zustandsvektor muß stetig sein

$$\bar{\boldsymbol{a}}_{k+1}^*(0) = \bar{\boldsymbol{a}}_k^*(h_k) = \bar{\boldsymbol{a}}^* \qquad (k = 1, 2, \ldots, N-1) \,. \tag{5.11}$$

In Verbindung mit der Übertragungsgleichung (5.7) lauten also die Übergangsbedingungen

$$\bar{\boldsymbol{a}}_k^* = \boldsymbol{T}_k^* \, \bar{\boldsymbol{a}}_{k-1}^* \qquad (k = 1, 2, \ldots, N) \,. \tag{5.12}$$

Die Elimination der Zwischengrößen liefert

$$\bar{\boldsymbol{a}}_N^* = \boldsymbol{T}_N^* \cdots \boldsymbol{T}_1^* \, \bar{\boldsymbol{a}}_0^* = \boldsymbol{T}^* \, \bar{\boldsymbol{a}}_0^* \tag{5.13}$$

mit $\boldsymbol{T}^*$ als Gesamt-Übertragungsmatrix. Auf Grund der beiden Randbedingungen — Vorgabe von σ_{zz} oder u_z am Rand 0 und am Rand N — lassen sich aus (5.13) die nicht vorgegebenen Komponenten von $\bar{\boldsymbol{a}}_0^*$ und $\bar{\boldsymbol{a}}_N^*$ berechnen, womit gemäß (5.12) alle reduzierten Zustandsvektoren an den Schichträndern ermittelt werden können. Die vollständigen „Anfangs"-Zustandsvektoren ergeben sich sodann nach (5.5) und der vollständige (transformierte) Spannungs- und Verschiebungszustand der Schicht k folgt mit (3.39) und (3.40) zu

$$\bar{\boldsymbol{a}}_k(z_k^*) = \boldsymbol{T}_k(z_k^*) \, \boldsymbol{C}_k \, \bar{\boldsymbol{a}}_k^*(0) \,, \tag{5.14}$$

$$\bar{\boldsymbol{b}}_k(z_k^*) = \bar{\boldsymbol{B}}_k \, \boldsymbol{T}_k(z_k^*) \, \boldsymbol{C}_k \, \bar{\boldsymbol{a}}_k^*(0) \,. \tag{5.15}$$

Es läßt sich übrigens — wie im einzelnen in [1] und [3] gezeigt wird — auch eine geschlossene Darstellung für den reduzierten Zustandsvektor $\bar{\boldsymbol{a}}_k^*$ angeben.

6. Sonderfälle

6.1. Der axialsymmetrische Verzerrungszustand

Hier besteht keine Abhängigkeit von der Umfangskoordinate φ: $f(\varphi) = \text{const.} = f$. Folglich gilt nach (3.1)

$$\hat{f} = \begin{cases} \sqrt{2\pi} f \\ 0 \end{cases} \quad \text{und} \quad \hat{\hat{f}} = \begin{cases} -\,i\sqrt{2\pi} f & \text{für} \quad n = 0 \\ 0 & \phantom{\text{für}} \quad n \neq 0 \end{cases} \tag{6.1}$$

und es folgt gemäß (3.8) und (3.9)

$$\hat{\boldsymbol{a}} = \sqrt{2\pi} \, \{\sigma_{zz}, \sigma_{zr}, -i\,\sigma_{z\varphi}, -i\,u_\varphi^*, u_r^*, -u_z^*\}^T \,, \tag{6.2}$$

$$\hat{\boldsymbol{b}} = \sqrt{2\pi} \, \{\sigma_{rr} + \sigma_{\varphi\varphi}, \sigma_{rr} - \sigma_{\varphi\varphi}, -i\,2\,\sigma_{r\varphi}\}^T \,. \tag{6.3}$$

Bei Beachtung dieses Zusammenhanges erhält man über (3.14), (3.28) bzw. (3.15), (3.29)

$$\bar{\boldsymbol{a}} = \boldsymbol{H} \boldsymbol{R} \, \hat{\boldsymbol{a}} = \sqrt{2\pi} \, \boldsymbol{Q} \, \bar{\boldsymbol{u}} \,, \tag{6.4}$$

$$\bar{\boldsymbol{b}} = \boldsymbol{H}' \boldsymbol{R}' \, \hat{\boldsymbol{b}} = \sqrt{2\pi} \, \boldsymbol{P} \, \bar{\boldsymbol{v}} \tag{6.5}$$

$$
\boldsymbol{U}_k(z_k^*) = \frac{c(z_k^*)}{2\,(1-\nu_k)} \begin{bmatrix}
2\,(1-\nu_k) - \lambda^* z_k^*\, t(z_k^*) & -\lambda^* z_k^* - (1-2\,\nu_k)\, t(z_k^*) & 0 & 0 & -\dfrac{E_k^*}{1+\nu_k}\lambda^{*2} z_k^*\, t(z_k^*) & \dfrac{E_k^*}{1+\nu_k}\lambda^*[\lambda^* z_k^* - t(z_k^*)] \\
\lambda^* z_k^* - (1-2\,\nu_k)\, t(z_k^*) & 2\,(1-\nu_k) + \lambda^* z_k^*\, t(z_k^*) & 0 & 0 & \dfrac{E_k^*}{1+\nu_k}\lambda^*[\lambda^* z_k^* + t(z_k^*)] & \\
0 & 0 & 2\,(1-\nu_k) & \dfrac{1-\nu_k}{1+\nu_k} E_k^* \lambda^*\, t(z_k^*) & & \\
0 & 0 & \dfrac{4\,(1-\nu_k^2)}{E_k^*\lambda^*}\cdot t(z_k^*) & & & \\
\dfrac{1+\nu_k}{E_k^*}\cdot z_k^*\, t(z_k^*) & \dfrac{1+\nu_k}{E_k^*}\left[z_k^* + \dfrac{(3-4\,\nu_k)}{\lambda^*} t(z_k^*)\right] & & \text{symmetrisch} & & \\
\dfrac{1+\nu_k}{E_k^*}\left[z_k^* - \dfrac{(3-4\,\nu_k)}{\lambda^*} t(z_k^*)\right] & & & & &
\end{bmatrix}
\tag{6.12}
$$

$$
{}^{(0)}\boldsymbol{U}_k(z_k^*) = \begin{bmatrix}
1 & -\lambda^* z_k^* & 0 & 0 & -\dfrac{E_k^*}{2\,(1-\nu_k^2)}\lambda^{*3} z_k^{*2} & \dfrac{E_k^*}{6\,(1-\nu_k^2)}\lambda^{*4} z_k^{*3} \\
\dfrac{\nu_k}{1-\nu_k}\lambda^* z_k^* & 1 & 0 & 0 & \dfrac{E_k^*}{1-\nu_k^2}\lambda^{*2} z_k^* & \\
0 & 0 & 1 & \dfrac{E_k^*}{2\,(1+\nu_k)}\lambda^{*2} z_k^* & & \\
0 & 0 & \dfrac{2\,(1+\nu_k)}{E_k^*} z_k^* & & & \\
\dfrac{1+\nu_k}{2\,(1-\nu_k)}\dfrac{\lambda^*}{E_k^*} z_k^{*2} & \dfrac{2\,(1+\nu_k)}{E_k^*} z_k^* & & \text{symmetrisch} & & \\
-\dfrac{(1+\nu_k)(1-2\,\nu_k)}{(1-\nu_k)\,E_k^*} z_k^* & & & & &
\end{bmatrix}.
\tag{6.13}
$$

mit

$$\boldsymbol{Q} = \begin{bmatrix} 1 & & & & & \\ & 1 & -i & & & \\ & -1 & -i & & & \\ & & & -i & -1 & \\ & & & -i & 1 & \\ & & & & & 2 \end{bmatrix} \quad \text{und} \quad \boldsymbol{P} = \begin{bmatrix} 1 & & \\ & 1 & -1 \\ & 1 & i \end{bmatrix} \tag{6.6), (6.7}$$

sowie

$$\overline{\boldsymbol{u}} = \{\overset{0}{\overline{\sigma}_{zz}},\ \overset{1}{\overline{\sigma}_{zr}},\ \overset{1}{\overline{\sigma}_{z\varphi}},\ \overset{1}{\overline{u}^*_{\varphi}},\ \overset{1}{\overline{u}^*_{r}},\ -\overset{0}{\overline{u}^*_{z}}\}^T \tag{6.8}$$

und

$$\overline{\boldsymbol{v}} = \{\overset{0}{\overline{\sigma_{rr}+\sigma_{\varphi\varphi}}},\ \overset{2}{\overline{\sigma_{rr}-\sigma_{\varphi\varphi}}},\ \overset{2}{\overline{2\,\sigma_{r\varphi}}}\}^T\,. \tag{6.9}$$

Mit (6.4) geht die Übertragungsgleichung (3.39) über in

$$\overline{\boldsymbol{u}}_k(z_k^*) = \boldsymbol{Q}^{-1}\,\boldsymbol{T}_k(z_k^*)\,\boldsymbol{Q}\,\overline{\boldsymbol{u}}_k(0) = \boldsymbol{U}_k(z_k^*)\,\overline{\boldsymbol{u}}_k(0) \tag{6.10}$$

und mit (6.5) Gl. (3.40) in

$$\overline{\boldsymbol{v}}_k(z_k^*) = \boldsymbol{P}^{-1}\,\overline{\boldsymbol{B}}_k\,\boldsymbol{Q}\,\overline{\boldsymbol{u}}_k(z_k^*) = \overline{\boldsymbol{V}}_k\,\overline{\boldsymbol{u}}_k(z_k^*)\,. \tag{6.11}$$

Die Ausrechnung liefert dann die neuen Matrizen $\boldsymbol{U}_k$ gemäß (6.12) bzw. ${}^{(0)}\boldsymbol{U}_k$ gemäß (6.13) und $\overline{\boldsymbol{V}}_k$ gemäß (6.14):

$$\overline{\boldsymbol{V}}_k = \begin{bmatrix} \dfrac{2\,\nu_k}{1-\nu_k} & 0 & 0 & 0 & \dfrac{\lambda^*}{1-\nu_k}E_k^* & 0 \\ 0 & 0 & 0 & 0 & -\dfrac{\lambda^*}{1+\nu_k}E_k^* & 0 \\ 0 & 0 & 0 & -\dfrac{\lambda^*}{1+\nu_k}E_k^* & 0 & 0 \end{bmatrix}. \tag{6.14}$$

Ein Blick auf die Matrix (6.12) lehrt, daß ein entkoppeltes Problem vorliegt: Der durch die Variablen $\{\sigma_{zz}, \sigma_{zr}, u_r^*, u_z^*\}$ beschriebene Zustand entspricht der axialsymmetrischen Beanspruchung ohne Torsion, der durch die Variablen $\{\sigma_{z\varphi}, u_\varphi^*\}$ beschriebene Zustand entspricht der reinen axialsymmetrischen Torsion. Beide Fälle wurden schon früher vom Verfasser behandelt [3], [9], [10].

Die Beziehung für den Halbraum als Schicht N lautet

$$\boldsymbol{X}_N\,\overline{\boldsymbol{u}}_{N-1} = 0 \tag{6.15}$$

mit

$$\boldsymbol{X}_N = \begin{bmatrix} 2\,(1-\nu_N) & -(1-2\,\nu_N) & 0 & 0 & 0 & -\dfrac{E_N^*}{1+\nu_N}\lambda^* \\ -(1-2\,\nu_N) & 2\,(1-\nu_N) & 0 & 0 & \dfrac{E_N^*}{1+\nu_N}\lambda^* & 0 \\ 0 & 0 & 2 & \dfrac{E_N^*}{1+\nu_N}\lambda^* & 0 & 0 \end{bmatrix}. \tag{6.16}$$

Bei *glatten Schichten* geht man analog zu Ziff. 5 vor und arbeitet mit dem reduzierten Zustandsvektor

$$\overline{\boldsymbol{u}}^* = \{\overset{0}{\overline{\sigma}_{zz}},\ \overset{0}{\overline{u}^*_{z}}\} \tag{6.17}$$

und der reduzierten Übertragungsmatrix

$$\boldsymbol{U}_k^* = \frac{c_k}{(\lambda^*\,h_k^* + t_k)} \begin{bmatrix} \dfrac{\lambda^*\,h_k^*}{c_k^2} + t_k & \dfrac{\lambda^*\,E_k^*}{2\,(1-\nu_k^2)}\left[t_k^2 - \left(\dfrac{\lambda^*\,h_k^*}{c_k}\right)^2\right] \\ \dfrac{2\,(1-\nu_k^2)}{\lambda^*\,E_k^*}\,t_k^2 & \dfrac{\lambda^*\,h_k^*}{c_k^2} + t_k \end{bmatrix} \tag{6.18}$$

bzw. — im Fall einer dünnen Schicht —

$${}^{(0)}\boldsymbol{U}_k^* = \begin{bmatrix} 1 & \dfrac{(\lambda^*\,h_k^*)^4}{12\,(1-\nu_k^2)}\,\dfrac{E_k^*}{h_k^*} \\ (1-\nu_k^2)\,\dfrac{h_k^*}{E_k^*} & 1 \end{bmatrix}. \tag{6.19}$$

Man bemerkt die Übereinstimmung von (6.18) und (6.19) mit (5.9) und (5.10).

Der Zusammenhang zwischen dem reduzierten und dem vollständigen Zustandsvektor an der Stelle $z_k = 0$ der Schicht k ist gegeben durch

$$\bar{u}_k(0) = W_k \, \bar{u}_k^*(0) \tag{6.20}$$

mit

$$W_k = \begin{bmatrix} 1 & 0 \\ 0 & 0 \\ 0 & 0 \\ 0 & 0 \\ -\dfrac{1+\nu_k}{E_k^* \lambda^*} \dfrac{\lambda^* h_k^* - (1-2\nu_k) t_k}{\lambda^* h_k^* + t_k} & -\dfrac{\lambda^* h_k^* t_k}{\lambda^* h_k^* + t_k} \\ 0 & -1 \end{bmatrix} . \tag{6.21}$$

Die für den axialsymmetrischen Verzerrungszustand maßgebenden Matrizen U_k, $\bar{V}_k$, X_N, U_k^* und W_k stimmen formal mit jenen des als Sonderfall in [1] behandelten zweidimensionalen Verzerrungszustandes (bezüglich der kartesischen Koordinaten x, z) überein. Diese Analogie ist in [3] für ein querisotropes Material dargestellt.

6.2. Schubbeanspruchung in einer Richtung ($|n| = 1$)

Unterliegt der Rand 0 einer Schubbelastung der Form

$$\sigma_{zx_0} = \tau(r) , \qquad \sigma_{zy_0} = 0 ,$$

wie es näherungsweise beim Blockieren des Rades eines Fahrzeugs zutrifft, so gilt für die Schubspannungskomponenten $\sigma_{zr_0} = -\tau(r) \cos\varphi$ und $\sigma_{z\varphi_0} = \tau(r) \cdot \sin\varphi$. Gemäß (3.1) folgt dann

$$\hat{\sigma}_{zr_0} = -\sqrt{\frac{\pi}{2}} \tau(r) \qquad \text{für} \quad |n| = 1 , \quad \text{sonst } 0$$

$$\hat{\hat{\sigma}}_{z\varphi_0} = \sqrt{\frac{\pi}{2}} \tau(r) \qquad \text{für} \quad |n| = 1 , \quad \text{sonst } 0 .$$

Daher wird nach (3.14)

$$\tau_{\mathrm{I}_0} = 0 \qquad \text{für alle } n$$

$$\tau_{\mathrm{II}_0} = -\sqrt{2\pi}\, \tau(r) \qquad \text{für } |n| = 1 , \quad \text{sonst } 0 .$$

Mithin ist die Bedeutung des Falles $|n| = 1$ erkannt.

Literatur

1 BUFLER, H., Theory of elasticity of a multilayered medium, J. of Elasticity **1**, S. 125 (1971).
2 KUO, J. T., Static response of a multilayered medium under inclined surface loads, J. Geoph. Res. **74**, S. 3195 (1969).
3 BUFLER, H., Spannungs- und Stabilitätsproblem eines elastisch-querisotropen Schichtsystems, Ing.-Arch. **41**, S. 278 (1972).
4 SNEDDON, I. N., Fouriertransforms, New York, Toronto, London (1951).
5 MUKI, T., Asymmetric problems of the theory of elasticity for a semi-infinite solid and a thick plate, in "Progress in Solid Mechanics" Vol. 1, ed. by I. N. Sneddon and R. Hill, North-Holland, Amsterdam and Interscience, New York (1960).
6 KOLCHIN, G. B., FAVERMAN, Z. A., The theory of elasticity of non-homogeneous bodies (bibliographical index of native and foreign literature), "Shtiintsa" publishing house, Kishinev (1972).
7 PESTEL, E. C. and LECKIE, F. A., Matrix methods in elastomechanics, New York, San Francisco, Toronto, London (1963).
8 MARGUERRE, K. und UHRIG, R., Berechnung vielgliedriger Schwingerketten, ZAMM **41**, S. 1 (1964).
9 BUFLER, H., Die Bestimmung des Spannungs- und Verschiebungszustandes eines geschichteten Körpers mit Hilfe von Übertragungsmatrizen, Ing.-Arch. **31**, S. 229 (1962).
10 BUFLER, H., Die Torsion der inhomogenen dicken Platte, ZAMM **43**, S. 389 (1963).

Eingereicht am 22. 3. 1973

Anschrift: Prof. Dr.-Ing. H. BUFLER, Institut für Mechanik (Bauwesen) der Universität Stuttgart, 7 Stuttgart 1, Keplerstraße 11, BRD

KLEINE MITTEILUNGEN

ZAMM 54, 119 – 121 (1974)

S. Y. Ruo

Slowly Oscillating Pointed Low Aspect Ratio Wings at Sonic Speed with Thickness Effect

1. Introduction

Wings in unsteady transonic flow have been extensively treated in [1] and [2] in which analyses are based on the linearized unsteady transonic small perturbation potential equation and the linearized tangency condition. The frequency of oscillation is assumed to be high and the thickness effect, a steady state property, is completely uncoupled from the unsteady equation.

The incorporation of the thickness effect in the unsteady flow analysis has been of interest to many investigators. The introduction of the parabolic method [3] for steady body of revolution at sonic speed has facilitated a method to restore some of the thickness effect in the unsteady flow analysis as can be seen in [4], [5] for two-dimensional flow and in [6] for the body of revolution. In this note, an approach is presented using the concepts of the parabolic method [3], the equivalence rule [7] and an iteration method [8] to approximately incorporate the thickness effect in the analysis of a slowly oscillating low aspect ratio, pointed wing at sonic speed.

2. Analysis

A possible form of the unsteady transonic flow equation capable of approximately incorporating the wing thickness effect is [4], [5], [6]

$$(1 - M_\infty^2)\,\hat\varphi_{xx} + \hat\varphi_{yy} + \hat\varphi_{zz} - 2\,M_\infty^2\,\hat\varphi_{xt} - M_\infty^2\,\hat\varphi_{tt} = M_\infty^2(\gamma + 1)\,\Phi_{xx}\,\hat\varphi_x\,, \tag{1}$$

where $\hat\varphi$ and Φ are respectively the unsteady and steady small perturbation potentials, and the former is considered to be only a small disturbance to the latter. M_∞ designates the free stream Mach number and γ is the ratio of specific heats. The right hand side of Eq. (1) is the coupling term of the steady part in unsteady flow. The term $M_\infty^2(\gamma+1)\,\Phi_x\hat\varphi_{xx}$ is neglected in the right hand side of Eq. (1) based on the argument that $\hat\varphi$ is a small disturbance to Φ and the contribution of this term is smaller than that of the term retained [5]. The coordinate system used is shown in Fig. 1.

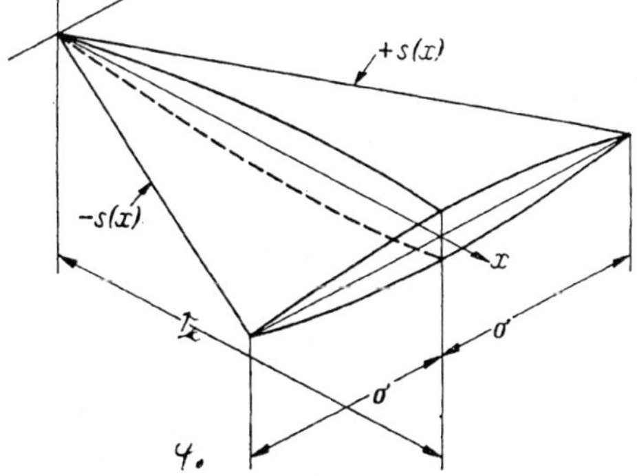

Fig. 1. Wing geometry

Following the parabolic method introduced in [3], one can reduce Eq. (1) for a wing performing low amplitude harmonic oscillation at $M_\infty = 1$ to

$$\varphi_{yy} + \varphi_{zz} - A\,\varphi_x + k^2\,\varphi = 0\,, \tag{2}$$

where φ is the amplitude of $\hat\varphi$ and k is the reduced frequency,

$$A = \Gamma + 2\,i\,k \quad \text{and} \quad \Gamma = (\gamma + 1)\,\Phi_{xx}\,.$$

Applying the Laplace transformation to Eq. (2), one gets

$$\bar\varphi_{yy} + \bar\varphi_{zz} - \lambda^2\,\bar\varphi = 0\,, \tag{3}$$

where $\bar\varphi$ is the Laplace transform of φ and $\lambda^2 = A\,p - k^2$. The dipole solution of Eq. (3), which vanishes at infinity, is

$$\bar\varphi(p, y, z) = \frac{1}{\pi}\frac{\partial}{\partial z}\int_{-s}^{+s}\bar\varphi(p, \eta, 0^+)\,K_0(\lambda\,r)\,d\eta \tag{4}$$

where $r = \sqrt{(y-\eta)^2 + z^2}$ and K_0 is the modified Bessel function of the second kind, of order zero.

For a low aspect ratio wing at low frequency, only the lower order terms of $(\lambda\,r)$ in K_0 need to be retained. After taking the inverse transformation of Eq. (4) term by term, one obtains the solution in the physical plane:

$$\begin{aligned}\varphi &= \frac{z}{\pi}\left\{I_0 + \frac{1}{4}\Lambda\,[G\,I_1 + 2\,I_3 + K\,II_1 - II_2] + \right.\\ &\quad \left. + \frac{1}{32}\Lambda^2\left[\left(G - \frac{3}{2}\right)I_3 + 2\,I_6 + K\,II_3 - II_4\right]\right\}\\ &= \frac{z}{\pi}\left\{I_0 + \frac{i\,k}{2}D\left[G\,I_1 + 2\,I_5 - \frac{2\,i\,k}{A}D\,II_1\right] + \right.\\ &\quad \left. + \frac{k^2}{8}D^2\left[-\left(G - \frac{3}{2}\right)I_3 - 2\,I_6 + \frac{2\,i\,k}{A}D\,II_3\right]\right\},\end{aligned} \tag{5}$$

where

$$G = \ln\left(\frac{1}{4}A\right) + 2\,c - 1\,,$$

Euler's constant $c = 0.5772\ldots$

$$D = \left(1 - \frac{i\,\Gamma}{2\,k}\right)\frac{\partial}{\partial x} + \frac{i\,k}{2}\,, \qquad \Lambda = 2\,i\,k\,D$$

$$E = \exp\,[K\,(x - \xi)]\,\ln\,[(x - \xi) + c]\,, \qquad K = \frac{k^2}{A}$$

$$I_0 = \int \frac{1}{r^2}\,\varphi\,d\eta\,, \quad I_1 = \int \varphi\,d\eta\,, \qquad I_3 = \int r^2\,\varphi\,d\eta$$

$$II_1 = \int E\,I_1\,d\xi\,, \quad II_3 = \int E\,I_3\,d\xi\,, \quad II_2 = \int E \int \varphi_x\,d\eta\,d\xi$$

$$I_5 = \int (\ln r)\,\varphi\,d\eta\,, \qquad II_4 = \int E \int r^2\,\varphi_x\,d\eta\,d\xi$$

$$I_6 = \int (r^2 \ln r)\,\varphi\,d\eta\,,$$

and the ranges of integration are expressed as $0 \leqq \xi \leqq x$ and $-s \leqq \eta \leqq +s$, throughout this note.

After differentiating Eq. (5) with respect to z and letting $z \to 0^+$, one gets the following:

$$\begin{aligned}\varphi_z(x, y, 0^+) &= \frac{1}{\pi}\left\{I_0 + \frac{i\,k}{2}D\left[G\,I_1 + 2\,I_5 + \frac{2\,i\,k}{A}D\,II_1\right] + \right.\\ &\quad \left. + \frac{k^2}{8}D^2\left[-\left(G - \frac{3}{2}\right)I_3 - 2\,I_6 + \frac{2\,i\,k}{A}D\,II_3\right]\right\}.\end{aligned} \tag{6}$$

For $\Gamma = 0$, Eq. (6) reduces to Eq. (3.7), page 33 of [1]. For a pointed wing

$$I_0 = -\oint \frac{\varphi_\eta}{y - \eta}\,d\eta\,,$$

and for a low frequency of oscillation

$$\exp\,[K\,(x - \xi)] \approx 1 + K\,(x - \xi)\,, \tag{7}$$

one may rewrite Eq. (6) as

$$\varphi_z = -\frac{1}{\pi}\,I_0 + \zeta[\varphi]\,, \tag{8}$$

where

$$\zeta[\varphi] = \frac{A}{4\pi}\left\{\frac{\partial}{\partial x}[N\,I_1 + 2\,I_7] - \frac{\partial^2}{\partial x^2}[II_{11}]\right\} - $$
$$- \frac{k^2}{4\pi}\left\{N\,I_1 + 2\,I_7 - 2\frac{\partial}{\partial x}[II_{11}] + \frac{\partial^2}{\partial x^2}[II_{12}]\right\}, \qquad (9)$$

$$N = \frac{1}{2}\ln\left(k^2 + \frac{\Gamma^2}{4}\right) + \ln\frac{1}{2} + c - 1 + i\left[\frac{\pi}{2} - \tan^{-1}\left(\frac{\Gamma}{2\,k}\right)\right],$$

$$I_7 = \int \varphi \ln|y-\eta|\,d\eta\,,\quad II_{11} = \int [\ln(x-\xi)]\,I_1\,d\xi\,,$$

$$II_{12} = \int (x-\xi)\,[\ln(x-\xi)]\,I_1\,d\xi\,.$$

Eq. (8) is generally applicable to a low aspect ratio, thin pointed wing with an unswept trailing edge and a smooth leading edge performing low amplitude and low frequency harmonic oscillations at sonic speed.

3. Iteration method

For a very low frequency case where $k/\Gamma \ll 1$, Eq. (9) reduces to

$$\zeta[\varphi] = \frac{\Gamma}{4\pi}\left\{\frac{\partial}{\partial x}[G_1\,I_1 + 2\,I_7] - \frac{\partial^2}{\partial x^2}[II_{11}]\right\} +$$
$$+ \frac{i\,k}{2\pi}\left\{\frac{\partial}{\partial x}[(G_1+1)\,I_1 + 2\,I_7] - \frac{\partial^2}{\partial x^2}[II_{11}]\right\}, \qquad (10)$$

where $G_1 = \ln\left(\frac{\Gamma}{4}\right) + c - 1$.

Following the iteration method described in [5], one may write

$$\varphi = \varphi^{(1)} + \varphi^{(2)} + \varphi^{(3)} + \cdots, \qquad (11)$$

where $\varphi^{(1)}$ is the slender wing (Munk-Jones) term and the remainder consists of higher order terms. Substituting Eq. (11) into the right hand side of Eq. (8) and equating the like order terms, one obtains the following relationships:

$$\varphi_z = -\frac{1}{\pi}\oint\frac{\varphi_\eta^{(1)}}{y-\eta}\,d\eta\,,\quad 0 = -\frac{1}{\pi}\oint\frac{\varphi_\eta^{(2)}}{y-\eta}\,d\eta + \zeta^{(1)}[\varphi^{(1)}]$$

$$0 = -\frac{1}{\pi}\oint\frac{\varphi_\eta^{(3)}}{y-\eta}\,d\eta + \zeta^{(1)}[\varphi^{(2)}] + \zeta^{(2)}[\varphi^{(1)}]\,,\quad \text{etc.,}$$

which can be written as

$$w^{(n)} = -\frac{1}{\pi}\oint\frac{\varphi_\eta^{(n)}}{y-\eta}\,d\eta\,, \qquad (12)$$

where

$$w^{(1)} = \varphi_z\,,\qquad w^{(2)} = -\zeta^{(1)}[\varphi^{(1)}]\,,$$
$$w^{(3)} = -\zeta^{(1)}[\varphi^{(2)}] - \zeta^{(2)}[\varphi^{(1)}]\,,\quad \text{etc.} \qquad (13)$$

It is assumed that the leading edge of the wing is pointed, i.e., $\varphi^{(n)}(x, \pm s, 0^+) = 0$, and the solution is given by

$$\varphi_y^{(n)} = \frac{1}{\pi\sqrt{s^2-y^2}}\oint\frac{w^{(n)}\sqrt{s^2-\eta^2}}{y-\eta}\,d\eta\,. \qquad (14)$$

With $w^{(n)}$ being given, in terms of the lower order solutions as shown in Eq. (13), $\varphi^{(n)}$ can be derived from Eq. (14).

In the subsequent analysis, only the first two terms on the right hand side of Eq. (11) will be considered for the case of a delta wing whose leading edge is expressed as $\pm s = \sigma\,x$.

Let the displacement of the mean surface of the wing, performing an oscillatory motion in a mode j with a zero mean angle of attack, be given by

$$z = \delta_j\,f_j(x, y)\,e^{i\,k\,t}\,,$$

and let the linearized tangency condition be written as

$$\varphi_{zj}(x, y, 0^+) = \delta_j\,[f_{xj} + i\,k\,f_j]\,. \qquad (15)$$

After the substitution of Eq. (15) into Eq. (12), with Eq. (14), one obtains the following equations for plunging and pitching oscillations:

(a) plunging: $f_1 = 1$

$$\varphi_1^{(1)}(x, y, 0^+) = -\,i\,k\,\delta_1\sqrt{s^2-y^2}$$
$$\varphi_1^{(2)}(x, y, 0^+) = -\frac{i\,k\,\delta_1\,\sigma\,s\,\Gamma}{4}(v + \ln x)\sqrt{s^2-y^2} \qquad (16)$$

(b) pitching: $f_2 = x - a$

$$\varphi_2^{(1)}(x, y, 0^+) = -\,\delta_2\,[1 + i\,k\,(x-a)]\sqrt{s^2-y^2}$$
$$\varphi_2^{(2)}(x, y, 0^+) = -\frac{\delta_2\,\sigma\,s}{8}\Big\{2\,\Gamma(v+\ln x) + 4\,i\,k(v+1+\ln x) +$$
$$+ i\,k\,\Gamma\left[x\left(3\,v + 3\ln x + \frac{5}{6} + \frac{2}{3}\frac{y^2}{s^2}\right) -\right.$$
$$\left.- 2\,a\,(v + \ln x)\right]\Big\}\sqrt{s^2+y^2}\,, \qquad (17)$$

where $v = c + \ln\frac{\Gamma\sigma^2}{16}$, and a is the arbitrary pitching axis location measured from the nose of the delta wing.

With the relationships given in Eqs. (16) and (17), the unsteady pressure distribution and the generalized aerodynamic force coefficients [1], up to the second order, can be obtained. In addition, the stability derivatives can be evaluated from these relationships.

4. Damping-in-pitch for a delta wing

Following Landahl's notation [1], the damping-in-pitch up to second order is expressed in terms of the generalized aerodynamic force coefficients as

$$C_{m_q} + C_{m\dot{\alpha}} = \frac{1}{k}\{[\bar{L}_{22}'']^{(1)} + [\bar{L}_{22}'']^{(2)}\}\,, \qquad (18)$$

where

$$[\bar{L}_{22}'']^{(n)} = \text{Imaginary part of}\left\{\frac{4}{S\,\delta_2}\iint_S f_2\,[\varphi_{x_2}^{(n)} + i\,k\,\varphi_2^{(n)}]\,dx\,dy\right\},$$

and S designates the area of the wing planform.

After substituting Eq. (17) into Eq. (18), one obtains the damping-in-pitch for a delta wing with leading edge $\pm s = \sigma\,x$ as follows:

$$\frac{2}{\pi\,AR}(C_{m_q} + C_{m\dot{\alpha}}) = -(1-a)^2 - \frac{AR^2}{64}\left\{\frac{1}{2}\left(3\,v + \frac{13}{4}\right) + \right.$$
$$\left.+ \frac{\Gamma}{5}\left(7\,v + \frac{21}{10}\right) - a\left[2\,(v+1) + \frac{\Gamma}{2}(5\,v+1) - a\,\Gamma\,v\right]\right\}, \qquad (19)$$

where AR denotes the aspect ratio. The first parentheses on the right hand side of Eq. (19) is the contribution from slender wing theory and the remainder results from the next higher order term provided by the present method. From Eq. (19) one observes that the damping-in-pitch for a delta wing, at the very low frequency range, remains nearly constant; thus the damping-in-pitch at the limit as $k \to 0$ can be calculated from the present approach.

The numerical results obtained using Eq. (19) for aspect ratio 1.0 and 3.0 delta wings of different thickness, as well as the results from Eq. (3.38) on page 37 of [1], are plotted in Figs. 2 and 3. The value of Γ is selected from that given in [6] using the method proposed in [4] for an equivalent ogive body ($R = \varepsilon\,x\,(2-x)$) having a cross-sectional area distribution along the x-axis identical to that for the wing under consideration (Fig. 1). Note that in the figures the thickness of the wing is expressed in term of the thickness ratio (ε) of the equivalent body of revolution.

Based on the results of the present method along with that of [1], as shown in Figs. 2 and 3, the contribution of the second and higher order terms to damping-in-pitch appears to have a destabilizing effect for $0 < a < 0.75$, and a stabilizing effect for $0.75 \leq a \leq 1.0$, in comparison with that for the slender wing theory. It is interesting to note that at $a \approx 0.75$ the second and higher order term contribution becomes nearly the same for the different thicknesses at very low frequency (present method), or for

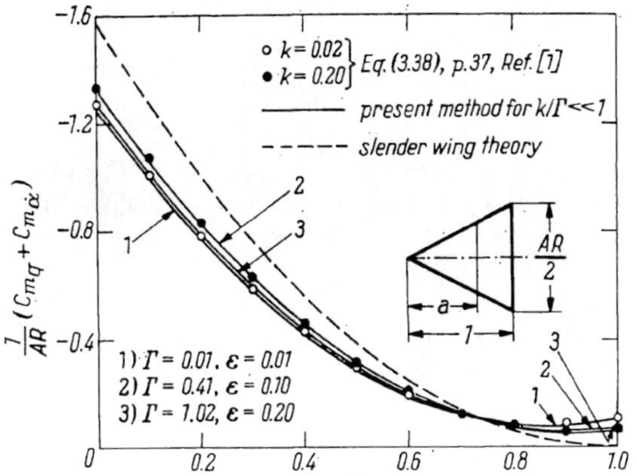

Fig. 2. Damping-in-pitch for an AR = 1.0 delta wing at very low frequency

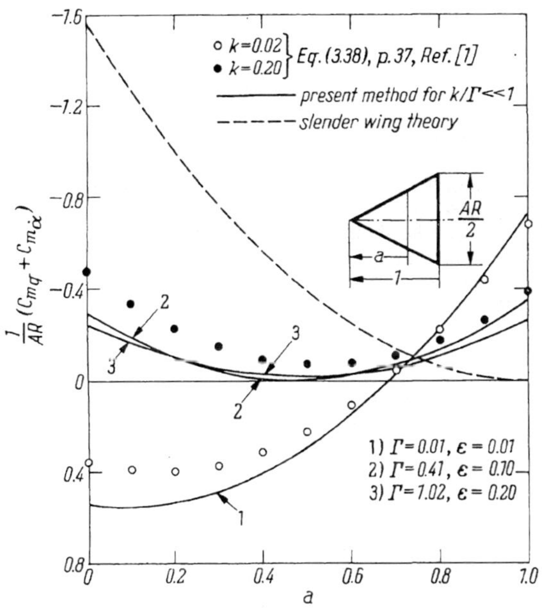

Fig. 3. Damping-in-pitch for an AR = 3.0 delta wing at very low frequency

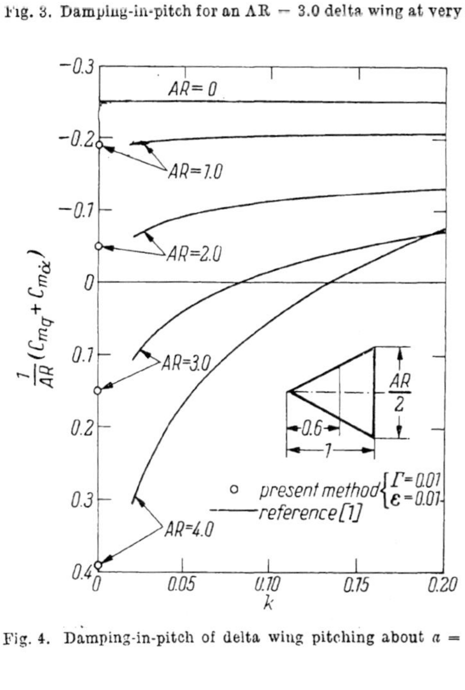

Fig. 4. Damping-in-pitch of delta wing pitching about $a = 0.6$

the different frequencies without thickness effect (method of [1]), and the cross-over value remains near 0.1 of the aspect ratio. It also shows that the thickness effect yields the same influence trend as that of the frequency on damping-in-pitch, and the influence of aspect ratio is far more pronounced than that of either frequency or thickness.

Plotted in Fig. 4 are the results from [1] and the present method for very thin delta wings of different aspect ratio pitching at an axis $a = 0.6$. It shows that the present method can be used to provide the case for $k \to 0$ where the linearized theory fails.

5. Concluding remarks

A closed-form solution for unsteady transonic flow, capable of taking the wing thickness into account in an approximate manner, has been presented. The numerical results show that the thickness has the same influence as the frequency on the damping-in-pitch, and that for moderately thin wings (ε: 0.10 $\sim$ 0.20) the thickness effect is not very significant. The effect of aspect ratio, however, is more pronounced than that of either thickness or frequency.

In this note only the very low frequency case has been considered. The moderate frequency case may be obtained by retaining more terms in Eq. (7); for this case, however, the closed-form solution is not easily attainable and one has to resort to a numerical method for evaluating the singular integrals. For a very high frequency case ($\Gamma/k \ll 1$) a closed-form solution is again possible.

In order to consider a higher aspect ratio wing, one needs to retain additional higher order terms of (λr) in $K_0(\lambda r)$ of Eq. (4). Also, the inclusion of $\varphi^{(3)}, \varphi^{(4)}, \ldots$ etc. in Eq. (11) allows investigation of other higher order term contributions by repeating the iteration technique.

The selection of a Γ value based on the equivalent body of revolution is not totally adequate because it is considered to depend only on the total cross-sectional area distribution along the x-axis, but not on the shape of the cross-sectional profile in the x and y directions of the wing. This deficiency can be improved upon by devising a technique to modify the value of Γ according to the wing cross-sectional profiles.

References

1 Landahl, M. T., Unsteady Transonic Flow, Pergamon Press, New York, 1961.
2 Davies, D. E., Three Dimensional Sonic Theory, AGARD Manual on Aeroelasticity, Part II – Aerodynamic Aspect, Chapter 4.
3 Oswatitsch, K. and Keune, F., The Flow Around Bodies of Revolution at Mach Number 1, Proceedings of the Conference on High-Speed Aeronautics, Polytechnic Institute of Brooklyn, Brooklyn, New York, 1955, pp. 113–131.
4 Hosokawa, I., A Simplified Analysis for Transonic Flows Around Thin Bodies, IUTAM Symposium Transsonicum, Springer-Verlag, Berlin, 1964, pp. 184–199.
5 Teipel, I., Die Instationären Luftkräfte bei der Machzahl 1, Zeitschrift für Flugwissenschaften, 12, pp. 6–14 (1964).
6 Ruo, S. Y. and Liu, D. D., Calculation of Stability Derivatives for Slowly Oscillating Bodies of Revolution at Mach 1.0, Lockheed Missiles and Space Co., LMSC/HREC D162375, 1971.
7 Oswatitsch, K., Similarity and Equivalence in Compressible Flow, Advances in Applied Mechanics, VI, Academic Press, New York, 1960, pp. 153–271.
8 Adams, M. C. and Sears, W. R., Slender-Body Theory – Review and Extension, J. Aero. Sci., 20, pp. 85–98 (1953).

Eingereicht am 1. 1. 1973

Anschrift: Dr. S. Y. Ruo, Lockheed-Georgia Company, Marietta, Georgia 30060, USA

ZAMM 54, 121–123 (1974)

E. Eder

Eine einfache Approximation für das vollständige elliptische Normalintegral erster und zweiter Gattung

1. Einleitung

Der Ausdruck

$$A(k) = \frac{k^2\,[\pi + 2\,\mathsf{K}(k)]}{8\,[\mathsf{E}(k) - (1 - k^2)\,\mathsf{K}(k)]} \tag{1}$$

$\mathsf{K}(k)$ = vollständiges (LEGENDREsches) elliptisches Normalintegral erster Gattung mit dem Modul k,
$\mathsf{E}(k)$ = vollständiges (LEGENDREsches) elliptisches Normalintegral zweiter Gattung mit dem Modul k,
unterscheidet sich — für nicht zu nahe an 1 liegende k-Werte — nur wenig von 1. Dies ermöglicht die Herleitung einer einfachen Approximation $\tilde{\mathsf{K}}(k)$ für $\mathsf{K}(k)$. $\tilde{\mathsf{K}}(k)$ ergibt eine Approximation $\tilde{\mathsf{E}}(k)$ für $\mathsf{E}(k)$. Mit den Approximationen $\tilde{\mathsf{K}}(k)$ und $\tilde{\mathsf{E}}(k)$ können u. a. Approximationen

$$\tilde{L}_n(a) = \int_0^a k^n \tilde{\mathsf{K}}(k)\, dk \quad \text{und} \quad \tilde{M}_n(a) = \int_0^a k^n \tilde{\mathsf{E}}(k)\, dk$$

für die Integrale

$$L_n(a) = \int_0^a k^n \mathsf{K}(k)\, dk \quad \text{und} \quad M_n(a) = \int_0^a k^n \mathsf{E}(k)\, dk$$

$(n = 0, 1, 2, \ldots)$ erhalten werden.

2. Approximation $\tilde{\mathsf{K}}(k)$

Es ist [1]

$$\mathsf{E}(k) - (1 - k^2)\, \mathsf{K}(k) = k\,(1 - k^2) \frac{d\mathsf{K}(k)}{dk}. \tag{2}$$

Setzt man in (1) $A(k) = 1$, so erhält man eine Näherung für $\mathsf{E}(k) - (1 - k^2)\, \mathsf{K}(k)$. Ersetzt man die linke Seite von (2) durch diese Näherung, dann erhält man die folgende Differentialgleichung für die Approximation $\tilde{\mathsf{K}}(k)$:

$$(1 - k^2) \frac{d\tilde{\mathsf{K}}(k)}{dk} - \frac{k}{4} \tilde{\mathsf{K}}(k) - \frac{\pi k}{8} = 0\,. \tag{3}$$

Die allgemeine Lösung von (3) lautet

$$\tilde{\mathsf{K}}(k) = C\,(1 - k^2)^{-1/8} - \frac{\pi}{2}.$$

Man erhält — wenn man C so festlegt, daß $\tilde{\mathsf{K}}(0) = \mathsf{K}(0) = \pi/2$ ist — für $\tilde{\mathsf{K}}(k)$ den Ausdruck

$$\tilde{\mathsf{K}}(k) = \pi\,[(1 - k^2)^{-1/8} - 0{,}5] = \pi\,(k'^{-1/4} - 0{,}5)\,, \tag{4}$$

$k' = \sqrt{1 - k^2}$ = komplementärer Modul.

Für zwischen 0 und 0,93 liegende k-Werte weicht $\tilde{\mathsf{K}}(k)$ von $\mathsf{K}(k)$ dem Betrage nach um weniger als 1% ab; für zwischen 0 und 0,5 liegende k-Werte ist die Abweichung dem Betrage nach maximal 4/1000%.

3. Approximation $\tilde{\mathsf{E}}(k)$

Es ist [2]

$$k\,(1 - k^2) \frac{d\mathsf{K}(k)}{dk} = \int k\, \mathsf{K}(k)\, dk\,. \tag{5}$$

(5) und (2) ergeben

$$\mathsf{E}(k) = (1 - k^2)\, \mathsf{K}(k) + \int k\, \mathsf{K}(k)\, dk\,. \tag{6}$$

Ersetzt man in (6) $\mathsf{K}(k)$ durch $\tilde{\mathsf{K}}(k)$ nach (4) und führt die Integration durch, so erhält man die folgende Approximation $\tilde{\mathsf{E}}(k)$ für $\mathsf{E}(k)$:

$$\tilde{\mathsf{E}}(k) = \frac{\pi}{14} + \frac{\pi}{4} k^2 + \frac{3\pi}{7} (1 - k^2)^{7/8} = \frac{9\pi}{28} - \frac{\pi}{4} k'^2 + \frac{3\pi}{7} k'^{7/4}\,. \tag{7}$$

Dabei wurde die Integrationskonstante gemäß $\tilde{\mathsf{E}}(0) = \mathsf{E}(0) = \pi/2$ festgelegt.

Für zwischen 0 und 1 liegende k-Werte weicht $\tilde{\mathsf{E}}(k)$ von $\mathsf{E}(k)$ dem Betrage nach um weniger als 1% ab; für zwischen 0 und 0,5 liegende k-Werte ist die Abweichung dem Betrage nach maximal 4/1000%.

4. Approximationen $\tilde{L}_n(a)$ und $\tilde{M}_n(a)$

Es ist nach (4) und (7)

$$\tilde{L}_n(a) = -\frac{\pi\, a^{n+1}}{2\,(n+1)} + \pi \int_0^a \frac{k^n\, dk}{(1 - k^2)^{1/8}} = -\frac{\pi\, a^{n+1}}{2\,(n+1)} + \pi\, I_n(a) \tag{8}$$

und

$$\tilde{M}_n(a) = \frac{\pi\, a^{n+1}}{14\,(n+1)} + \frac{\pi\, a^{n+3}}{4\,(n+3)} + \frac{3\pi}{7} \int_0^a \frac{k^n\, dk}{(1 - k^2)^{1/8}} - \frac{3\pi}{7} \int_0^a \frac{k^{n+2}\, dk}{(1 - k^2)^{1/8}} = \frac{\pi\, a^{n+1}}{14\,(n+1)} + \frac{\pi\, a^{n+3}}{4\,(n+3)} + \frac{3\pi}{7} [I_n(a) - I_{n+2}(a)]\,. \tag{9}$$

Bei der Ermittlung der Approximationen $\tilde{L}_n(a)$ und $\tilde{M}_n(a)$ sind demnach Integrale

$$I_i(a) = \int_0^a \frac{k^i\, dk}{(1 - k^2)^{1/8}} \qquad (i = 1, 2, \ldots)$$

zu berechnen. Die Substitution $x = \arccos(k/a)$ ergibt

$$I_i(a) = a^{i+1} \int_0^{\pi/2} \frac{\sin x \cos^i x\, dx}{(1 - a^2 \cos^2 x)^{1/8}}\,.$$

Man erhält [1], S. 103,

$$I_i(a) = \frac{a^{i+1}}{2} B\left(\frac{i+1}{2}, 1\right) F\left(\frac{i+1}{2}, \frac{1}{8}, \frac{i+3}{2}; a^2\right), \tag{10}$$

B = Betafunktion oder EULERsches Integral erster Gattung
F = Hypergeometrische Reihe.

Ist i ungerade, dann ist $I_i(a)$ eine elementare Funktion von a. Daher sind $\tilde{L}_n(a)$ und $\tilde{M}_n(a)$ für ungerades n elementare Funktionen von a.

Insbesondere ergibt (10) für $a = 1$ [2], S. 8,

$$I_i(1) = \frac{\Gamma\left(\frac{7}{8}\right) \Gamma\left(\frac{i+1}{2}\right)}{2\, \Gamma\left(\frac{4i+11}{8}\right)}, \tag{11}$$

Γ = Gammafunktion oder EULERsches Integral zweiter Gattung. (11) ergibt mit (8) und (9)

$$\tilde{L}_n(1) = \int_0^1 k^n \tilde{\mathsf{K}}(k)\, dk = -\frac{\pi}{2\,(n+1)} + \frac{\pi\, \Gamma\left(\frac{7}{8}\right) \Gamma\left(\frac{n+1}{2}\right)}{2\, \Gamma\left(\frac{4n+11}{8}\right)},$$

$$\tilde{M}_n(1) = \int_0^1 k^n \tilde{\mathsf{E}}(k)\, dk = \frac{\pi}{14\,(n+1)} + \frac{\pi}{4\,(n+3)} + \frac{3\pi\, \Gamma\left(\frac{7}{8}\right) \Gamma\left(\frac{n+1}{2}\right)}{2\,(4n+11)\, \Gamma\left(\frac{4n+11}{8}\right)}.$$

Gemäß (6) erhält man

$$L_1(a) = \int_0^a k\, \mathsf{K}(k)\, dk = \mathsf{E}(a) - (1 - a^2)\, \mathsf{K}(a)\,.$$

Andererseits ergibt (10) und (8)

$$\tilde{L}_1(a) = \int_0^a k\, \tilde{\mathsf{K}}(k)\, dk = \frac{4\pi}{7} - \frac{\pi\, a^2}{4} - \frac{4\pi}{7} (1 - a^2)^{7/8}\,.$$

Für zwischen 0 und 1 liegende a-Werte weicht $\tilde{L}_1(a)$ von $L_1(a)$ dem Betrage nach um weniger als 1% ab; für zwischen

0 und 0,96 liegende a-Werte ist die Abweichung dem Betrage nach maximal 0,3%.

5. Kennzeichnung der Güte der Approximationen

Im folgenden wird die Güte der Approximationen durch Angabe der ersten Glieder der Entwicklungen nach Potenzen von $k^2/8$ bzw. $a^2/8$ deutlich gemacht:

$$A(k) - 1 = \frac{3}{2}\left(\frac{k^2}{8}\right)^2 + 11\left(\frac{k^2}{8}\right)^3 + \cdots, \; k < 1,$$

$$\mathsf{K}(k) - \widetilde{\mathsf{K}}(k) = -\frac{\pi}{2}\left(\frac{k^2}{8}\right)^3 - \frac{25\pi}{4}\left(\frac{k^2}{8}\right)^4 - \cdots, \quad k < 1,$$

$$\mathsf{E}(k) - \widetilde{\mathsf{E}}(k) = -\frac{\pi}{2}\left(\frac{k^2}{8}\right)^3 - \frac{11\pi}{4}\left(\frac{k^2}{8}\right)^4 - \cdots, \quad k < 1,$$

$$L_n(a) - \widetilde{L}_n(a) = -\frac{\pi a^{n+1}}{2(n+7)}\left(\frac{a^2}{8}\right)^3 - \frac{25\pi a^{n+1}}{4(n+9)}\left(\frac{a^2}{8}\right)^4 - \cdots, \quad a < 1, \quad n = 0, 1, 2, \ldots,$$

$$M_n(a) - \widetilde{M}_n(a) = -\frac{\pi a^{n+1}}{2(n+7)}\left(\frac{a^2}{8}\right)^3 - \frac{11\pi a^{n+1}}{4(n+9)}\left(\frac{a^2}{8}\right)^4 - \cdots, \quad a < 1, \quad n = 0, 1, 2, \ldots$$

Literatur

1 GRÖBNER, W. und HOFREITER, N., Integraltafeln, Bd. II, Springer-Verlag, Wien 1966, S. 40.

2 MAGNUS, W. und OBERHETTINGER, F., Formeln und Sätze für die speziellen Funktionen der mathematischen Physik, Springer-Verlag, Berlin 1943, S. 106.

Eingegangen am 28. 2. 1973

Anschrift: Wiss. Oberrat Dr. ERICH EDER, Inst. f. Mechanische Technologie II u. Betriebstechnik d. TH-Wien, A 1040 Wien, Karlsplatz 13, Österreich

ZAMM 54, 123 – 125 (1974)

H. BRASS

Eine Restdarstellung bei Gregoryschen Quadraturverfahren ungerader Ordnung

1. Das Resultat

Das GREGORYsche Quadraturverfahren der Ordnung r besteht darin, $\int_a^b f(x)\,dx$ anzunähern durch

$$Q_n[f] := h\left\{\sum_{\nu=0}^{n} f(x_\nu) - \sum_{\nu=0}^{r} L_{\nu+1}\left[\Delta^\nu f(x_{n-\nu}) + (-1)^\nu \Delta^\nu f(x_0)\right]\right\}. \tag{1}$$

Hierbei ist

$$h := \frac{b-a}{n}, \quad x_\nu := a + \nu h,$$

$$L_\nu := (-1)^{\nu-1}\int_0^1 \binom{x}{\nu}\,dx$$

und die Differenzen sind zur Spanne h zu nehmen.

Die GREGORY-Verfahren haben gegenüber anderen Quadraturverfahren eine Reihe von Vorzügen (MARTENSEN [4], LÖTZBEYER [3]), die eine genauere Untersuchung ihres Fehlers

$$R_n[f] := \int_a^b f(x)\,dx - Q_n[f]$$

wünschenswert erscheinen lassen. Während man im Fall gerader Ordnung eine Restdarstellung von der „klassischen" Struktur $R_n[f] = \text{const}\, f^{(r+2)}(\xi)$ $(\xi \in [a, b]$, ebenso im folgenden) erhält (BARRET [1], MARTENSEN [4]), ist der Fall ungerader Ordnung komplizierter. MARTENSEN ([4], [5]) hat für ihn mehrere Restdarstellungen hergeleitet, es sei genannt

$$R_n[f] = -2L_{r+2}h^{r+2}\left[f^{(r+1)}(a) + \frac{b-a}{2} f^{(r+2)}(\xi)\right].$$

Das Ziel dieser Note ist der Beweis einer weiteren Restdarstellung, die — mindestens für große n — engere Restschranken liefert als die bisher bekannten, weil sie das asymptotische Verhalten des Fehlers, nämlich

$$R_n[f] = -L_{r+2}h^{r+2}[f^{(r+1)}(a) + f^{(r+1)}(b)] + O(h^{r+3})$$

richtig wiedergibt. Es handelt sich um den folgenden

Satz: *Für das Gregory-Verfahren der ungeraden Ordnung r gilt die folgende Restdarstellung, sofern f eine stetige $(r+3)$-te Ableitung hat:*

$$R_n[f] = -L_{r+2}h^{r+2}[f^{(r+1)}(a) + f^{(r+1)}(b)] + c_r h^{r+3} f^{(r+3)}(\xi)$$

mit

$$c_r = (b-a)\left(\frac{r+1}{2}L_{r+2} - L_{r+3}\right) - h\left(2L_{r+4} - (r+2)L_{r+3} + \frac{3r^2+7r+4}{12}L_{r+2}\right).$$

Dieser Satz ist eine direkte Konsequenz der beiden folgenden Lemmata

Lemma 1: *Es gilt*

$$\int_{x_{n-r}}^{x_n} f(x)\,dx - h\left\{\sum_{\nu=n-r}^{n} f(x_\nu) - \sum_{\nu=0}^{r} L_{\nu+1}\left[\Delta^\nu f(x_{n-\nu}) + (-1)^\nu \Delta^\nu f(x_{n-r})\right]\right\}$$
$$= -L_{r+2}h^{r+2}[f^{(r+1)}(x_{n-r}) + f^{(r+1)}(x_n)] + h^{r+4}\tilde{c}_r f^{(r+3)}(\xi).$$

Hierbei ist

$$\tilde{c}_r := -2L_{r+4} + 2L_{r+3} + \frac{3r^2-r-4}{12}L_{r+2} > 0.$$

Lemma 2: *Es gilt*

$$\int_{x_\mu}^{x_{\mu+1}} f(x)\,dx - h\left\{f(x_\mu) + \sum_{\nu=0}^{r}(-1)^\nu L_{\nu+1}\Delta^{\nu+1}f(x_\mu)\right\}$$
$$= L_{r+2}h^{r+2}[f^{(r+1)}(x_{\mu+1}) - f^{(r+1)}(x_\mu)] + h^{r+4}\left(\frac{r+1}{2}L_{r+2} - L_{r+3}\right)f^{(r+3)}(\xi).$$

Der Koeffizient von $f^{(r+3)}(\xi)$ ist positiv.

Um den Satz herzuleiten, hat man nur Lemma 2 für $\mu = 0, 1, \ldots, n-r-1$ anzuschreiben und alle diese Formeln zu Lemma 1 zu addieren. Wenn man noch beachtet, daß alle Summanden mit $f^{(r+3)}(\xi)$ zu einem zusammengezogen werden können, da sie alle positive Koeffizienten haben, entsteht so direkt die Behauptung.

2. Beweis von Lemma 1

Die Formel

$$\int_{x_0}^{x_r} g(x)\,dx - h\left\{\sum_{\nu=0}^{r} g(x_\nu) - \sum_{\nu=0}^{r} L_{\nu+1}\left[\Delta^\nu g(x_{r-\nu}) + (-1)^\nu \Delta^\nu g(x_0)\right]\right\}$$
$$= -L_{r+2}h^{r+2}[g^{(r+1)}(x_0) + g^{(r+1)}(x_r)] + h^{r+4}\tilde{c}_r g^{(r+3)}(\xi) \tag{2}$$

ist äquivalent mit Lemma 1, denn sie wird mittels $g(x) = f(x + x_{n-r} - x_0)$ in Lemma 1 übergeführt. Nun zeigt der Vergleich mit (1) sofort, daß hier auf der linken Seite $R_r[g]$ steht. Im Spezialfall $n = r$ ist die GREGORYsche Quadraturformel mit der von NEWTON-COTES identisch (siehe

MARTENSEN [4]), so daß (2) eine Aussage über den Rest der NEWTON-COTES-Formel ist. Für diesen gilt bekanntlich die PEANOsche Darstellung (siehe KRYLOV [2], S. 89):

$$R_r[g] = \int_{x_0}^{x_r} g^{(r+1)}(x)\, K(x)\, dx$$

wo K eine nichtpositive, bezüglich $m := (x_0 + x_r)/2$ gerade Funktion ist. Partielle Integration unter Verwendung der Stammfunktionen

$$\overline{K}(x) = \int_m^x K(y)\, dy\,, \quad \overline{\overline{K}}(x) = \int_{x_0}^x \overline{K}(y)\, dy$$

ergibt

$$R_r[g] = \overline{K}(x_r)\, [g^{(r+1)}(x_r) + g^{(r+1)}(x_0)] + \int_{x_0}^{x_r} g^{(r+3)}(x)\, \overline{\overline{K}}(x)\, dx\,.$$

Aus den oben genannten Eigenschaften von K folgert man leicht $\overline{\overline{K}} \geqq 0$ und hat somit

$$R_r[g] = \overline{K}(x_r)\, [g^{r+1}(x_r) + g^{r+1}(x_0)] + g^{r+3}(\xi) \int_{x_0}^{x_r} \overline{\overline{K}}(x)\, dx\,. \quad (3)$$

Setzt man hier $g_0(x) := x^{r+1}$ ein und vergleicht mit der bekannten Restdarstellung der NEWTON-COTES-Formel (siehe MARTENSEN [4]

$$R_r[g] = -2\, L_{r+2}\, h^{r+2}\, g^{(r+1)}(\xi)\,,$$

so folgt sofort $\overline{K}(x_r) = -L_{r+2}\, h^{r+2}$. Um nun $\tilde{c}_r$ zu bestimmen, setzt man

$$g_1(x) := \frac{\prod_{\nu=0}^{r+2} (x - x_\nu)}{(r+3)!\; h^{r+3}} \quad (4)$$

in (3) ein. Die notwendige Kenntnis von $R_r[g_1]$ verschafft man sich in folgender Weise durch die GREGORYsche Quadraturformel: Man ersetzt in (1) r durch $r + 3$ und n durch r. Die entstandene GREGORY-Formel der Ordnung $r + 3$ integriert Polynome des Grades $r + 3$ exakt und läßt sich deuten als Summe aus $Q_r[f]$ und drei Korrekturtermen. Die Auswertung der Korrekturterme für g_1 ist nicht schwierig und ergibt $R_r[g_1]$, damit erhält man nach etwas Rechnung den oben angegebenen Wert für $\tilde{c}_r$.

3. Beweis von Lemma 2

Vorweg bemerken wir, daß es ausreicht, den Fall $\mu = 0$ zu beweisen, die allgemeinere Aussage des Lemmas geht daraus durch eine einfache Translation hervor.

Das Funktional

$$G[f] := f(x_0) + \sum_{\nu=0}^{r} (-1)^\nu L_{\nu+1}\, \Delta^{\nu+1} f(x_0)$$

kann offenbar auch in der Form

$$G[f] = \sum_{\nu=0}^{r+1} a_\nu f(x_\nu)$$

geschrieben werden. Die Beweisstruktur wird einfacher, wenn wir eine unten benötigte Aussage über a_0 als Lemma vorwegschicken:

Lemma 3: *Für alle $r \geqq 3$ ist $(r + 1)\, a_0 > 1$.*

Beweis: Durch Vergleich der beiden Formen von G erhält man

$$a_0 = 1 - L_1 - L_2 \cdots - L_{r+1}\,.$$

Die Aussage wird also bewiesen sein, wenn gezeigt ist, daß die Folge

$$b_\nu := \nu \left(1 - \sum_{\mu=1}^{\nu} L_\mu\right)$$

monoton wächst, denn es gilt

$$b_3 = 3\left(1 - \frac{1}{2} - \frac{1}{12} - \frac{1}{24}\right) = \frac{9}{8} > 1\,.$$

Es ist

$$\begin{aligned} b_\nu - b_{\nu-1} &= 1 - \sum_{\mu=1}^{\nu-1} L_\mu - \nu\, L_\nu \\ &= 1 - \sum_{\mu=1}^{\nu-1} L_\mu - \nu \left(\frac{1}{\nu+1} - \sum_{\mu=1}^{\nu-1} \frac{L_\mu}{\nu - \mu + 1}\right) \\ &= \frac{1}{\nu+1} + \sum_{\mu=1}^{\nu-1} \frac{\mu - 1}{\nu - \mu + 1} L_\mu > 0\,, \end{aligned}$$

wobei von der bekannten Rekursionsformel (siehe STEFFENSEN [6], S. 106)

$$\sum_{\mu=1}^{\nu} \frac{L_\mu}{\nu - \mu + 1} = \frac{1}{\nu + 1}$$

Gebrauch gemacht wurde.

Der Beweis von Lemma 2 ergibt sich aus der von der durch Integration der NEWTONschen Interpolationsformel leicht zu erhaltenden Beziehung (siehe STEFFENSEN [6], S. 104)

$$\int_{x_0}^{x_1} f(x)\, dx - h\, G[f] = L_{r+2}\, h^{r+3} f^{(r+2)}(\xi)\,. \quad (5)$$

Die linke Seite dieser Gleichung kann man auch in der Form

$$\int_{x_0}^{x_{r+1}} f'(x)\, K_1(x)\, dx \quad (6)$$

mit einem leicht angebbaren K_1 schreiben. Wir benötigen nur

$$K_1(x)|_{[x_0, x_1]} = x_0 - x + a_0\, h\,. \quad (7)$$

(6) wird nun durch partielle Integration unter Einführung der Stammfunktionen

$$K_{\nu+1}(x) := \int_{x_0}^x K_\nu(y)\, dy\,, \quad \nu = 1, 2, \ldots, r + 1\,,$$

umgeformt. Hierbei verschwinden die ausintegrierten Terme, denn es ist

$$K_{\nu+1}(x_{r+1}) = \int_{x_0}^{x_{r+1}} K_\nu(y)\, dy = 0\,,$$

weil der untersuchte Ausdruck im Spezialfall $f_0(x) = x^\nu$ $(\nu < r + 2)$ Null wird, wie man aus (5) abliest. Man erhält also

$$\int_{x_0}^{x_1} f(x)\, dx - h\, G[f] = \int_{x_0}^{x_{r+1}} f^{(r+2)}(x)\, K_{r+2}(x)\, dx\,, \quad (8)$$

oder, unter Einführung von

$$K^*_{r+2}(x) = \begin{cases} K_{r+2}(x) - L_{r+2}\, h^{r+2} & \text{für } x \in [x_0, x_1] \\ K_{r+2}(x) & \text{für } x \in [x_1, x_{r+1}] \end{cases}$$

$$\int_{x_0}^{x_1} f(x)\, dx - h\, G[f] = L_{r+2}\, h^{r+2} [f^{(r+1)}(x_1) - f^{(r+1)}(x_0)] + \\ + \int_{x_0}^{x_{r+1}} f^{(r+2)}(x)\, K^*_{r+2}(x)\, dx\,.$$

Setzt man hier $f(x) = x^{r+2}$ ein und vergleicht mit (5), so erkennt man $\int_{x_0}^{x_{r+1}} K^*_{r+2}(x)\, dx = 0$. Somit ergibt eine weitere partielle Integration

$$\int_{x_0}^{x_1} f(x)\, dx - h\, G[f] = L_{r+2}\, h^{r+2} [f^{(r+1)}(x_1) - f^{(r+1)}(x_0)] - \\ - \int_{x_0}^{x_{r+1}} f^{(r+3)}(x)\, K_{r+3}(x)\, dx\,, \quad (9)$$

wobei

$$K_{r+3}(x) := \int_{x_0}^x K^*_{r+2}(y)\, dy$$

gesetzt ist. Unser nächstes Ziel ist der Nachweis von $K_{r+3}(x) \leqq 0$. Im Fall $r = 1$ hat man

$$K_{r+3}(x)|_{[x_0, x_1]} = -\frac{(x-x_0)^4}{24} + \frac{5\,h(x-x_0)^3}{72} - \frac{h^3(x-x_0)}{24},$$

$$K_{r+3}(x)|_{[x_1, x_2]} = \frac{h(x-x_2)^3}{72},$$

woraus die Behauptung direkt nachzuprüfen ist, im weiteren kann also $r > 1$ vorausgesetzt werden. Wegen

$$K_{r+3}(x_0) = K_{r+3}(x_{r+1}) = 0 \quad \text{und}$$

$$K^*_{r+2}(x_0) = -L_{r+2}\,h^{r+2} < 0$$

ist sicher $K_{r+3}(x) \leqq 0$, wenn K^*_{r+2} höchstens einen Zeichenwechsel hat. Letzteres sieht man so ein: Auf $[x_1, x_{r+1}]$ ist $K^*_{r+2} \equiv K_{r+2}$ und es ist $K_{r+2}(x) \geqq 0$, wie man durch Vergleich von (5) und (8) erkennt. Auf $[x_0, x_1]$ wächst K^*_{r+2} monoton, denn dort hat K_{r+1} positives Vorzeichen, weil (vgl. (7))

$$K_{r+1}(x)|_{[x_0, x_1]} = -\frac{(x-x_0)^{r+1}}{(r+1)!} + a_0\,h\,\frac{(x-x_0)^r}{r!}$$

$$= \frac{(x-x_0)^r}{(r+1)!}\,h\left[(r+1)\,a_0 - \frac{x-x_0}{h}\right] > 0$$

nach Lemma 3 ist.

Um nun den Beweis abzuschließen, hat man in (9) auf das letzte Integral den verallgemeinerten Mittelwertsatz anzuwenden und $\int_{x_0}^{x_{r+1}} K_{r+3}(x)\,dx$ durch die Einsetzung der Funktion (4) zu bestimmen.

Literatur

1 Barret, W, On the remainder term in numerical integration formulae, Journal London Math. Soc. 27, S. 465–470 (1952).
2 Krylov, V. J., Approximate calculation of integrals (translated by A. H. Stroud), Macmillan Co. New York 1962 (Übersetzung aus dem Russischen, Moskau 1959).
3 Lötzbeyer, W., Asymptotische Eigenschaften linearer und nichtlinearer Quadraturformeln, ZAMM 52, T 211 – T 214 (1972).
4 Martensen, E., Optimale Fehlerschranken für die Quadraturformel von Gregory. ZAMM 44, S. 159–168 (1964).
5 Martensen, E., Zur Restglieddarstellung der Gregoryschen Quadraturformel ungerader Ordnung, Numer. Math. 15, S. 229–233 (1970).
6 Steffensen, J. F., Interpolation, Chelsea Publishing Comp., New York 1950.

Eingereicht am 28. 2. 1973

Anschrift: Prof. Dr. H. Brass, Mathematisches Institut der T. U. Clausthal, 3392 Clausthal-Zellerfeld, Erzstr. 1, BRD

ZAMM 54, 125–126 (1974)

A. Buckley

A Note on Matrices A = I + H, H Skew-Symmetric*)

1. Introduction

We wish to consider a real matrix $A = I + H$ where H is skew-symmetric and I is the identity. We examine the decomposition $A = L\,U$, where L is lower triangular with diagonal elements 1 and U is upper triangular, show that the decomposition may always be obtained without pivoting, and give an a priori bound for the diagonal elements u_{ii} of U. We note that a bound on the spectral condition number of A is easily given. These results apply in particular to the case where the elements of H may be large, so that A is not diagonally dominant.

Such matrices arise naturally in the solution by finite differences of certain (possibly) nonlinear, hyperbolic partial differential equations. For example, take the equation

$$y_t + G(y, y_x, y_{xx}, \ldots) + \delta^2\, y_{xxx} = 0$$

*) Appears in Ph. D. thesis of the author, U. of British Columbia, Vancouver, Canada.

which is to be satisfied on the real line R. Let G be any smooth function. Assume initial condition $y(x, 0) = f(x)$, $x \in R$, and periodic boundary condition $y(x, t) = y(x + 1; t)$ for $x \in R$ and $t \geqq 0$ are given. Choosing a difference scheme as suggested by Kreiss [1] it is not difficult to show that the linear system $A\,x = (I + H)\,x = b$, where H is skew-symmetric and has the form

$$H = \begin{bmatrix} 0 & \times & \times & & & & & \times & \times \\ \times & 0 & \times & \times & & & & & \times \\ \times & \times & 0 & \times & \times & & & & \\ & \times & \times & 0 & \times & \times & & & \\ & & & & \ddots & & & & \\ \times & & & & & \times & \times & 0 & \times \\ \times & \times & & & & & \times & \times & 0 \end{bmatrix},$$

is obtained. (The elements in the corners are a result of the periodic boundary conditions.) Typically, the non-zero elements of H are larger than 1 in magnitude.

To numerically solve the differential equation, this system of linear equations must be solved repeatedly. It is therefore important to minimize the number of operations required to solve the system, and so it is desirable to obtain the $L\,U$ decomposition without pivoting. We must ensure then that L and U are defined and, ideally, we should give a bound on the elements of L and U. Unfortunately we are only able to give a bound on the u_{ii}, although computational experience of the author and of Sjöberg [4] with certain Toeplitz matrices H indicates that no growth of the off diagonal elements of L or U occurs.

2. The results

We recall a simple

Lemma: *The eigenvalues of the real $q \times q$ skew-symmetric matrix H are given by*

$$\pm\, i\,\lambda_j \quad j = 1, 2, \ldots, \frac{q}{2} \text{ if } q \text{ is even, or}$$

$$0,\ \pm\, i\,\lambda_j \quad j = 1, 2, \ldots, \frac{q-1}{2} \text{ if } q \text{ is odd.}$$

Each λ_j is a nonnegative eigenvalue of the hermitean matrix $i\,H$.

We then obtain the following

Theorem 1: *Let the real $q \times q$ matrix $A = I + H$, where H is skew-symmetric. Then A is nonsingular, the decomposition $A = L\,U$ exists (without pivoting) and the natural pivot elements u_{ii} of U are nonzero and bounded:*

$$1 \leqq u_{ii} \leqq 1 + \lambda^2, \quad i = 1, 2, \ldots, q;$$

λ is the largest nonnegative eigenvalue of $i\,H$.

Proof: For $k \leqq q$, let A_k denote the submatrix of A given by the first k rows and columns. Similarly define H_k, L_k and U_k. Recalling the triangular forms of L and U,

$$\det A_k = \det L_k \cdot \det U_k = \prod_{i=1}^{k} u_{ii}\,.$$

Hence

$$u_{kk} = \frac{\det A_k}{\det A_{k-1}}\,.$$

Assume k is even, and set $p = k/2$. (The odd case is similar.) Let $0 \leqq \lambda_1 \leqq \cdots \leqq \lambda_p$ denote the nonnegative eigenvalues of $i\,H_k$, and let $0 = \mu_0 \leqq \mu_1 \leqq \cdots \leqq \mu_{p-1}$ denote those of $i\,H_{k-1}$. By a well known theorem concerning eigenvalues of hermitean matrices (see, for example, Noble [3], p. 415),

$$0 \leqq \lambda_1 \leqq \mu_1 \leqq \lambda_2 \leqq \cdots \leqq \mu_{p-1} \leqq \lambda_p\,.$$

Hence, recalling that a determinant is given as the product of the eigenvalues, and noting that $A_k = I_k + H_k$ has, from the Lemma, eigenvalues $1 \pm i\,\lambda_j$, $j = 1, 2, \ldots, p$ (and similarly for A_{k-1}), we have that

$$u_{kk} = \frac{\prod_{j=1}^{p}(1+\lambda_j^2)}{\prod_{j=1}^{p-1}(1+\mu_j^2)} = (1+\lambda_p^2)\prod_{j=1}^{p-1}\frac{1+\lambda_j^2}{1+\mu_j^2} \leqq 1 + \lambda_p^2\,,$$

and

$$u_{kk} = (1 + \lambda_1^2) \prod_{j=1}^{p-1} \frac{1 + \lambda_{j+1}^2}{1 + \mu_j^2} \geqq 1 + \lambda_1^2 \geqq 1 \,.$$

The assertions of the theorem now follow.

Finally, we bound the spectral condition number $c(A)$ of A. Here $\|\cdot\|$ denotes the spectral norm of A.

Theorem 2: *Let the largest and smallest nonnegative eigenvalues of $i\,H$ be $\lambda_{\max}$ and $\lambda_{\min}$, respectively, where H is real and skew-symmetric. Then*

$$c(A) \equiv \|A\| \, \|A^{-1}\| = \left(\frac{1 + \lambda_{\max}^2}{1 + \lambda_{\min}^2}\right)^{1/2} \leqq (1 + \lambda_{\max}^2)^{1/2} \leqq$$

$$\leqq 1 + \lambda_{\max} \,.$$

Proof: It is well-known (see, for example, LANCASTER [2], p. 224) that

$$c(A) = \left(\frac{\alpha_n}{\alpha_1}\right)^{1/2}$$

where α_n and α_1 are the eigenvalues of $A^T A$ of largest and smallest modulus, respectively. But $A^T A = I - H^2$ since H is skew-symmetric. Using the Lemma then, $A^T A$ has eigenvalues $1 + \lambda_j^2$ and the result follows.

References

1 KREISS, H. O. and WIDLUND, O., Difference Approximations for Initial Value Problems for Partial Differential Equations, Report NR 7, Department of Computer Science, Uppsala University, Sweden, (Sept. 1967).

2 LANCASTER, P., Theory of Matrices, Academic Press, New York/London 1969, p. 224.

3 NOBLE, B., Applied Linear Algebra, Prentice Hall, Inc., Englewood Cliffs, New Jersey 1969, p. 415.

4 SJOBERG, A., Numerical Solution of the Korteweg-de Vries Equation, Report NR 25, Department of Computer Science, Uppsala University, Sweden, (Oct. 1969).

Eingereicht am 5. 4. 1973

Anschrift: A. BUCKLEY, Math. Dept. McGill University, Montreal 101, Quebec, Canada

ZAMM 54, 126 – 127 (1974)

J. BAUMGARTE*)

Zeittransformation zur analytischen Schrittregulierung bei der numerischen Integration gewöhnlicher Differentialgleichungen

Bei einer numerischen Integration von Bewegungsgleichungen treten in der Nähe von Singularitäten Schwierigkeiten auf. Derartige Singularitäten sind immer dann vorhanden, wenn bereits das Potential singulär ist. Die numerischen Schwierigkeiten können weitgehend vermieden werden, wenn man anstelle der physikalischen Zeit t auf eine neue unabhängige Variable s übergeht [1], [2], welche durch die Differentialgleichung

$$t' = \frac{dt}{ds} = \mu(x_i) \,, \qquad \mu > 0 \,, \tag{1}$$

definiert ist. Durch diese Zeittransformation kann man bei geeigneter Wahl von $\mu > 0$ eine *analytische Schrittanpassung* erreichen, welche die numerische Genauigkeit verbessert.

Die Zeittransformation legt auch die Transformation der LAGRANGE-Funktion fest. Wenn die Bewegungsgleichungen von N Massenpunkten in der physikalischen Zeit t durch die LAGRANGE-Funktion $L = L(x_i, \dot x_i, t)$ gegeben werden (die $3\,N$ Komponenten der N Ortsvektoren werden als ein $3\,N$-dimensionaler Vektor x_i interpretiert), dann werden sie in der „fiktiven Zeit" s durch die transformierte LAGRANGE-Funktion L^*

$$L^*(x_i, x'_i, t, t', E) = \mu\, L\left(x_i, \frac{x'_i}{\mu}, t\right) - E(t' - \mu) \,,$$

$$i = 1, 2, \ldots, 3\,N \,, \tag{2}$$

beschrieben. Hierbei bedeutet der LAGRANGEsche Multiplikator E die Gesamtenergie. Die Zeit t ist jetzt eine Koordinate. Liegt ein konservatives Problem vor, d. h. hängt L nicht explizit von t ab, dann wird die Gesamtenergie konstant, also $E = E_0 = \text{const.}$

Die Differentialgleichung der Zeittransformation (1) ist instabil. Es besteht jedoch die Möglichkeit, $\mu = \mu(x_i) > 0$, so zu wählen, daß im konservativen Fall folgendes erreicht wird:

a) Die Lösung der Differentialgleichung (1) $t = t(s, x_i, x'_i)$ kann direkt angegeben werden. Damit entfällt die Instabilität.

b) Die Zeittransformation (1) glättet die Singularitäten des Potentials. Dadurch wird ein besseres numerisches Verhalten der in der „fiktiven Zeit" s beschriebenen Bewegungsgleichung erreicht [3].

Zur Bestimmung dieser speziellen Funktion $\mu(x_i)$ geht man von (einem Viertel) der Spur des Trägheitstensors

$$J = \frac{1}{2} \sum_{(i)} m_i x_i^2 \,, \qquad i = 1, 2, \ldots, 3\,N \,, \tag{3}$$

aus. Durch Differenzieren folgt unter Benutzung der Bewegungsgleichungen und des Energiesatzes (mit U = Potential):

$$(\dot J)^\cdot = (\sum m_i x_i \dot x_i)^\cdot = \sum m_i \ddot x_i x_i + \sum m_i \dot x_i^2$$

$$= -\sum x_i \frac{\partial U}{\partial x_i} + 2\,(E_0 - U) \,. \tag{4}$$

Mit der Abkürzung

$$W(x_j) = 2\,U(x_j) + \sum_{(i)} x_i \frac{\partial U}{\partial x_i} \tag{5}$$

erhält man durch Integration:

$$\dot J = \sum m_i x_i \dot x_i = 2\,E_0 t + C - \int W\,dt \,. \tag{6}$$

C ist eine Integrationskonstante. Wir wählen nun

$$\mu(x_j) = \frac{k}{W(x_j)} \,, \tag{7}$$

wobei k eine beliebige Konstante ist, die aber dasselbe Vorzeichen wie W haben muß.

Es folgt

$$\sum m_i x_i \dot x_i = 2\,E_0 t + C - k\,s \tag{8}$$

oder

$$t = \frac{k\,s - C}{2\,E_0} + \frac{\sum m_i x_i x'_i}{2\,E_0\,\mu} \,. \tag{9}$$

Ist das Potential eine homogene Funktion des Vektors x_i, also

$$U(\alpha\, x_j) = \alpha^n\, U(x_j) \,, \tag{10}$$

wobei α eine beliebige Konstante ist und n den Grad der Homogenität angibt, dann folgt aus dem EULERschen Theorem:

$$\sum_{(i)} x_i \frac{\partial U}{\partial x_i} = n\, U(x_j) \,, \qquad \text{also} \quad W = (2 + n)\, U \,. \tag{11}$$

Beispiel 1: *Satellitenbewegung*

1.a) KEPLER-Bewegung

Mit K^2 = Gravitationsparameter und $r = (x_1^2 + x_2^2 + x_3^2)^{1/2}$ lautet das Potential

$$U = -\frac{m\,K^2}{r} \,. \tag{12}$$

Der Grad der Homogenität ist: $n = -1$. Damit wird $W = U = -\frac{m\,K^2}{r}$. Wir wählen $k = -m\,K^2$ und erhalten als Zeittransformation

$$t' = r \,. \tag{13}$$

*) Technische Universität Braunschweig, Mechanik-Zentrum, Lehrstuhl A für Mechanik.

Für eine elliptische Bewegung, mit $(-E_0) = h > 0$, bekommt man aus (9)

$$t = \frac{m K^2 s + C}{2h} - \frac{m \sum x_i x_i'}{2hr}. \tag{14}$$

Man kann zeigen, daß die Lösungen der Bewegungsgleichungen $x_i = x_i(s)$, welche die transformierte LAGRANGE-Funktion (2) liefert, stabil im Sinne von LJAPUNOV sind [2], [4]. Würde man jedoch zur Berechnung der Lösungen $x_i = x_i(t)$ nicht $t = t(s, x_i, x_i')$ gemäß (14) sondern die Differentialgleichung $t' = r$ benutzen, so wären dann die Lösungen $x_i(t)$ instabil.

1. b) Berücksichtigung des Erdwulstpotentials

In diesem Fall hat das Potential die Gestalt

$$U = -\frac{m K^2}{r} + m K^2 \sum_{(n)} c_n \frac{P_n(\cos\vartheta)}{r^{n+1}}, \quad n = 1, 2, \ldots, \tag{15}$$

wobei die c_n Konstante und die P_n die n-ten LEGRENDREschen Polynome sind. Es ist V_n homogen vom Grad $-(n+1)$. Wir wollen neben dem Zentral-Potential $-\frac{m K^2}{r}$ nur den dominierenden Anteil $m K^2 V_2$ des Erdwulst-Potentials betrachten. Aus dem EULERschen Theorem folgt:

$$\sum x_i \frac{\partial V_2}{\partial x_i} = -3 V_2. \tag{16}$$

Mit $k = -m K^2$ erhält man

$$W = -\frac{m K^2}{r} + m K^2 \left[2 V_2 + \sum x_i \frac{\partial V_2}{\partial x_i}\right]$$

$$= -m K^2 \left(\frac{1}{r} + V_2\right) \tag{17a}$$

und

$$t' = \frac{r}{1 + r V_2}. \tag{17b}$$

Damit tritt anstelle (9):

$$t = \frac{m K^2 s + C}{2h} - \frac{1 + r V_2}{2hr} m \sum x_i x_i'. \tag{18}$$

Beispiel 2: *Dreikörperproblem*

Es seien m_α, die Massen und $r_{\alpha\beta}$ die gegenseitigen Abstände der Körper. Das Gravitationspotential (Γ = Gravitationskonstante) beträgt

$$U = -\Gamma \sum_{\alpha<\beta} \frac{m_\alpha m_\beta}{r_{\alpha\beta}}. \tag{19}$$

Die 9 Komponenten der drei Ortsvektoren $\boldsymbol{r}_\alpha$ werden interpretiert als 9-dimensionaler Vektor x_i. Als Funktion der Ortsvektoren $\boldsymbol{r}_\alpha$ ist das Potential homogen vom Grad (-1), also

$$\sum_{(\alpha)} \left(\boldsymbol{r}_\alpha, \frac{\partial U}{\partial \boldsymbol{r}_\alpha}\right) = -U, \quad \alpha = 1, 2, 3. \tag{20}$$

Die Klammer in (20) bedeutet das Skalarprodukt. Damit wird auch

$$\sum_{(i)} x_i \frac{\partial U}{\partial x_i} = -U, \quad i = 1, 2, \ldots, 9. \tag{21}$$

Anstelle (6) erhält man

$$\sum m_i x_i \dot{x}_i = 2 E_0 t + C - \int U\, dt. \tag{22}$$

Zur Schrittanpassung wählen wir die Zeittransformation

$$t' = \frac{k}{U} > 0. \tag{23}$$

Damit erhält man für $t = t(s, x_i, x_i')$:

$$t = \frac{k s - C}{2 E_0} + \frac{U \sum m_i x_i x_i'}{2 E_0 k}, \quad \frac{k}{2 E_0} > 0. \tag{24}$$

Die Verallgemeinerung auf mehr als drei Körper ist leicht durchzuführen.

Die hier entwickelte Zeittransformation kann auch in allen gestörten Fällen angewandt werden, wenn man ein „Zeit-Element" [3] mit hinzunimmt.

Der Verfasser dankt der Kommission für Raumforschung der Schweizerischen Naturforschenden Gesellschaft für ihre Unterstützung.

Literatur

1 STIEFEL, E. L., and SCHEIFELE, G., Linear and regular celestial mechanics, Springer-Verlag, Berlin/Heidelberg/New York 1971.
2 BAUMGARTE, J., Numerical stabilization of the differential equations of Keplerian motion, Celestial Mechanics 5, 490–501 (1972), D. Reidel Publishing Company, Dordrecht – Holland.
3 BAUMGARTE, J., and STIEFEL, E., Examples of transformations improving the numerical accuracy of the integration of differential equations, erscheint in Celestial Mechanics.
4 SCHWARZ, H. R., Stability of Kepler motion, Computer Methods in Applied Mechanics and Engineering 1, 279–299 (1972), North-Holland Publishing Company.

Eingereicht am 11. 4. 1973

Anschrift: Prof. Dr.-Ing. JOACHIM BAUMGARTE, 334 Wolfenbüttel, Lärchenweg 1, BRD

ZAMM 54, 127–128 (1974)

T. ROLSKI/D. STOYAN

Two Classes of Semi-Orderings and their Application in the Queuing Theory

In the last time some papers have appeared in which semi-ordering relations of distribution functions (d. f.) are used to obtain bounds for characteristics of queuing systems. In the present paper we suggest two new classes of semi-ordering relations which are useful for the queuing theory and others areas.

We define $F_1 \leqq_{1.n} F_2$ for d.f.'s F_1 and F_2 such that

$$\int_{-\infty}^{0} |x|^n \, dF_i(x) < \infty; \quad i = 1, 2 \tag{1.1}$$

if for every real x

$$\int_{-\infty}^{x} \int_{-\infty}^{x_n} \cdots \int_{-\infty}^{x_2} F_1(x_1)\, dx_1 \cdots dx_n \geqq \int_{-\infty}^{x} \int_{-\infty}^{x_n} \cdots \int_{-\infty}^{x_2} F_2(x_1)\, dx_1 \cdots dx_n \tag{2.1}$$

Simultaneously we define $F_1 \leqq_{2.n} F_2$ for d.f.'s F_1 and F_2 with

$$\int_{0}^{\infty} x^n \, dF_i(x) < \infty; \quad i = 1, 2 \tag{1.2}$$

if for every real x

$$\int_{x}^{\infty} \int_{x_n}^{\infty} \cdots \int_{x_2}^{\infty} \bar{F}_1(x_1)\, dx_1 \cdots dx_n \leqq \int_{x}^{\infty} \int_{x_n}^{\infty} \cdots \int_{x_2}^{\infty} F_2(x_1)\, dx_1 \cdots dx_n \tag{2.2}$$

where $\bar{F}_i(x) = 1 - F_i(x)$.
In the case $n = 1$ the relations $\leqq_{n.1}$ and $\leqq_{n.2}$ are equivalent to the relations $\overset{(3)}{\leqq}$ and $\overset{(2)}{\leqq}$ in [3].

Obviously, we have

$$F_1 \leqq_{k.n} F_2 \Rightarrow F_1 \leqq_{k.n+1} F_2; \quad k = 1, 2.$$

By partial integration one can prove that

$$\int_{-\infty}^{\infty} \varphi(t)\, dF_1(t) \geqq \int_{-\infty}^{\infty} \varphi(t)\, dF_2(t) \tag{3.1}$$

holds (if the integrals are existing) for $F_1 \leqq_{1.n} F_2$ and for an arbitrary function φ of the form

$$\varphi(t) = \int_{t}^{\infty} \int_{x_n}^{\infty} \cdots \int_{x_2}^{\infty} G(x_1)\, dx_1 \cdots dx_n,$$

where G is increasing, right continuous and

$$\left|\int_0^\infty x^n\, dG(x)\right| < \infty .$$

For example, the functions $\varphi(t) = e^{-st}$, $s > 0$, are of this type. Analogously, if $F_1 \leqq_{2.n} F_2$,

$$\varphi(t) = \int_{-\infty}^{t} \int_{-\infty}^{x_n} \cdots \int_{-\infty}^{x_2} G(x_1)\, dx_1 \cdots dx_n ,$$

where G is nondecreasing, right continuous and

$$\int_{-\infty}^{0} |x|^n\, dG(x) < \infty ,$$

then

$$\int_{-\infty}^{\infty} \varphi(t)\, dF_1(t) \leqq \int_{-\infty}^{\infty} \varphi(t)\, dF_2(t) . \tag{3.2}$$

One can prove that the validity of (2.1) and (2.2) is equivalent to

$$\int_{-\infty}^{\infty} (x - t)_+^n\, dF_1(t) \geqq \int_{-\infty}^{\infty} (x - t)_+^n\, dF_2(t); \qquad -\infty < x < \infty$$

$$\int_{-\infty}^{\infty} (t - x)_+^n\, dF_1(t) \leqq \int_{-\infty}^{\infty} (t - x)_+^n\, dF_2(t); \qquad -\infty < x < \infty$$

where

$$z_+^n = \begin{cases} z^n; & z \geqq 0 \\ 0; & z < 0 . \end{cases}$$

Using this fact it can be proved the following property: If $F_1 \leqq_{1.n} F_2$ ($F_1 \leqq_{2.n} F_2$) and H is a d.f. with (1.1) (or (1.2)) then

$$F_1 * H \leqq_{1.n} F_2 * H \; (F_1 * H \leqq_{2.n} F_2 * H) . \tag{4}$$

By (4) we obtain

$$F_1 \leqq_{1.n} F_2 \Rightarrow F_1^{*(\nu)} \leqq_{1.n} F_2^{*(\nu)} \; (F_1 \leqq_{2.n} F_2 \Rightarrow F_1^{*(\nu)} \leqq_{2.n} F_2^{*(\nu)}),$$

$\nu = 1, 2, \ldots .$

Finally, $F_1 \leqq_{1.n} F_2$ ($F_1 \leqq_{2.n} F_2$) and

$$\int_{-\infty}^{\infty} |x|^n\, dF_i(x) < \infty ; \qquad i = 1, 2$$

implies

$$\int_{-\infty}^{\infty} x\, dF_1(x) \leqq \int_{-\infty}^{\infty} x\, dF_2(x) .$$

Sufficient condition for $\leqq_{1.n}$ ($\leqq_{2.n}$)

In the case of equality of the first n moments of d.f.'s F_1 and F_2 the relation $F_1 \leqq_{1.n} F_2$ ($F_1 \leqq_{2.n} F_2$) holds if $F_1 - F_2$ changes the sign n times and there exists a real number t_0 such that for every $i > t_0$

$F_1(t) - F_2(t) > 0$ if n is even (odd)

$F_1(t) - F_2(t) < 0$ if n is odd (even) .

This criterion follows from a theorem proved by KARLIN and NOVIKOFF [2].

Applications to the queuing theory

1) In many formulae in the queuing theory LAPLACE-STIELTJES transforms are occuring. The relations $\leqq_{1.n}$ are useful to comparing LAPLACE-STIELTJES transforms (and, consequently, queuing parameters) since

$$F_1 \leqq_{1.n} F_2 , \quad F_i(0) = 0; \qquad i = 1, 2$$

implies

$$\int_0^\infty e^{-st}\, dF_1(t) \geqq \int_0^\infty e^{-st}\, dF_2(t); \qquad s \geqq 0 . \tag{5}$$

For example, let us consider the machine interference problem (s machines, 1 operative). In case of exponential distribution (parameter: λ) of running times and general distribution function $R(t)$ of repair time, the average number $A(R)$ of machines running is equal to

$$A(R) = s \frac{\sum_{i=0}^{s-1} \binom{s-1}{i} c_i^{-1}}{1 + s r \lambda \sum_{i=0}^{s-1} \binom{s-1}{i} c_i^{-1}}$$

$$c_0 = 1, \; c_i = \prod_{j=0}^{i} \frac{\Phi(j\lambda)}{1 - \Phi(j\lambda)} .$$

$$\Phi(s) = \int_0^\infty e^{-st}\, dR(t) , \quad r = \int_0^\infty t\, dR(t) \qquad \text{(see [4])}.$$

Since $A(R)$ is monotonous in Φ, from $R_1 \leqq_{1.n} R_2$ follows $A(R_1) \geqq A(R_2)$.

2) The semi-invariants s_n of the stationary waiting time in $G\,I/G/1$ with d.f. of the difference between service and interarrival times equal $K(t)$ are given by

$$s_n = \sum_{\nu=1}^{\infty} \frac{1}{\nu} \int_0^\infty x^n\, dK^{*(\nu)}(x); \qquad n = 1, 2, \ldots$$

Let Σ_1 and Σ_2 be $G\,I/G/1$ systems with d.f.'s $K_1(t)$ and $K_2(t)$. Then, if $K_1 \leqq_{2.n} K_2$, for the corresponding semi-invariants $s_k^{(1)}$ and $s_k^{(2)}$ the inequality

$$s_k^{(1)} \leqq s_k^{(2)}; \qquad k \geqq n$$

holds. We remark that BERGMANN [1] proved a similar result using more complicated semi-ordering relations.

References

1 BERGMANN, R., Qualitätsunterschiede bei Wartemodellen vom Typ G/G/1. Math. Operationsforsch. Statist. 4, 5 (1974)
2 KARLIN, S. and NOVIKOFF, A., Generalized convex inequalities. Pacific J. Math. 13, 1951–1979 (1963).
3 STOYAN, D., Monotonieeigenschaften stochastischer Modelle. ZAMM 52, 23–34 (1972).
4 TAKACS, L., Introduction to the theory of queues. Oxford University Press, New York 1962.

Eingereicht am 13. 4. 1973

Anschriften: TOMASZ ROLSKI, Wroclaw University, Mathematical Institute, Poland;
Dr. DIETRICH STOYAN, 92 Freiberg (Sachs.), Tschaikowskistr. 85, DDR

BUCHBESPRECHUNGEN

K. Stange, Angewandte Statistik: Teil I: Eindimensionale Probleme. XVI + 592 S. m. 227 Abb. Berlin/Heidelberg/New York 1970. Springer-Verlag: Preis geb. DM 58,—. Teil II: Mehrdimensionale Probleme. XI + 505 S. m. 117 Abb. Berlin/Heidelberg/New York 1971. Springer-Verlag. Pr. geb. DM 49,—.

In den beiden vorliegenden Bänden, die zusammen fast 1100 Seiten umfassen, hat sich der Autor die Aufgabe gestellt, „Mathematiker der angewandten Richtung, Naturwissenschaftler, Ingenieure, Wirtschaftswissenschaftler und andere an Hand zahlreicher Anwendungen in das Wesen statistischen Denkens einzuführen". Eine Einteilung des Stoffes in ein- und mehrdimensionale Probleme ergibt die Trennung in die beiden Bände.

Im 1. Band werden zunächst die geläufigen Verfahren zur Auswertung von Meßreihen bezüglich eines Merkmals ausführlich dargestellt. Eine Einführung in die Grundbegriffe der Wahrscheinlichkeitsrechnung erfolgt vorwiegend an Hand von Beispielen. Dabei hätte man etwas genauer auf den wichtigen Begriff einer (international gebräuchlichen) Zufallsgröße, ihre Verteilungsfunktion und ihre Wahrscheinlichkeitsdichte eingehen können. Die Abschnitte 7 bis 11 enthalten statistische Methoden, bei denen normalverteilte Ausgangswerte zugrunde liegen, wie z. B. χ^2-Test, F-Test, Konfidenzintervalle für Mittelwerte, Varianz und Variationskoeffizient und Toleranzgrenzen. Die auf diskreten Verteilungen wie der Binomialverteilung, der POISSON-verteilung und der hypergeometrischen Verteilung beruhenden Testverfahren und Konfidenzschätzungen werden in den drei letzten Abschnitten des 1. Bandes behandelt, wobei der Autor auch auf den Zusammenhang der Verteilungen untereinander und auf ihre Verwendung in der statistischen Qualitätskontrolle eingeht.

Der 2. Band beinhaltet vor allem die Methoden der Korrelations- und Regressionsanalyse. Zunächst werden für zwei Merkmale Verfahren zur Auswertung vorliegenden Beobachtungsmaterials angegeben. Bei der Erläuterung des Modells der Korrelationsanalyse vermißt man eine genaue Definition der Begriffe Unabhängigkeit und Unkorreliertheit zweier Zufallsgrößen. An eine Herleitung der Verteilung des Korrelationskoeffizienten (als „Korrelationszahl" bezeichnet) schließen sich Testverfahren für den Korrelationskoeffizienten sowie Anwendungsbeispiele an. In den Abschnitten über lineare Regressionsanalyse für zwei Veränderliche werden neben den üblichen Test- und Schätzverfahren auch Nebenbedingungen und veränderliche Versuchsvarianz berücksichtigt sowie Toleranzgrenzen für die Zielgröße angegeben. Bei der mehrfachen linearen Regression liegt der Schwerpunkt auf Untersuchungen für dreidimensionale Zufallsgrößen. Abschnitte über Stichprobenverfahren, wo insbesondere das mehrstufige Modell, geschichtete Stichproben und Optimalitätskriterien bei der Auswahl einer Stichprobe erörtert werden, und die Monte-Carlo-Methode ergänzen den 2. Band. Ein Tabellenanhang enthält die wichtigsten Zahlentafeln, die für die praktische Lösung von Aufgaben erforderlich sind. Für eine Neuauflage wünschte man sich ein zusammenfassendes Literaturverzeichnis am Schluß des Buches.

Die Darstellung ist klar und anschaulich und für den Ingenieur gut lesbar. Die theoretischen Überlegungen basieren überwiegend auf einfachen mathematischen Hilfsmitteln, wie sie dem Studierenden der angewandten Wissenschaften zur Verfügung stehen. Einen breiten Raum nehmen die Anwendungen der statistischen Methoden auf praktische Probleme ein. Zahlreiche Beispiele, vorwiegend aus dem Bereich der Ingenieurwissenschaften, zeigen die Vielfalt der Anwendungsmöglichkeiten und geben dem in der Praxis tätigen Statistiker wertvolle Hinweise für die Bearbeitung statistischer Aufgaben.

Dresden R. STORM

J. C. Rotta, Turbulente Strömungen. (Leitfäden der angewandten Mathematik und Mechanik, Band 15). 267 S. m. 104 Fig. Stuttgart 1972. B. G. Teubner. Preis geb. 68,— DM.

Die kinetische Gastheorie und ihre mathematische Ordnung, die MAXWELL und BOLTZMANN durch Benutzen der Wahrscheinlichkeitsrechnung einführten, waren schon in der Pionierphase der Turbulenz, beim Ringen um eine erste Lösung Vorbild und Anregung. PRANDTLs und TAYLORs Begriffe über Wirbelballen und Mischungsweg sind analog zu den Begriffen Molekül und mittlere freie Weglänge der kinetischen Gastheorie aufgestellt. Tatsächlich ergaben auch die frühen experimentellen Erkundungen, daß z. B. die wahrscheinliche Dichteverteilung der Geschwindigkeitsschwankung hinter Maschengittern im Windkanal vorzüglich mit der GAUSSschen Normalverteilung übereinstimmt. Aber es zeigte sich bald, daß eine Theorie, die nur mit statistischen Mittelwerten operiert, nicht in der Lage war, unabhängig von Messungen die praktisch interessierenden Fragestellungen z. B. nach dem Intensitätsspektrum oder nach der Größe von Wirbelelementen und deren zeitlichem Abklingen in natürlicher Strömung zu beantworten. Die Theoretiker übernahmen bisher lediglich das Interpretieren von entsprechenden Meßergebnissen und versuchten, durch spezielle Vereinfachungen diese Ergebnisse mathematisch nachzuzeichnen. Der mathematische Aufwand, der im ganzen hierzu getrieben werden muß, ist recht anspruchsvoll. So wird es der Leser sehr begrüßen, daß der Verfasser nicht sogleich das spezielle physikalisch-technische Problem mit der mathematischen Theorie verwebt, sondern daß in den ersten Kapiteln abgetrennt die mathematischen Begriffe und Rechenverfahren sehr sauber dargestellt sind. In folgenden Kapiteln werden die Spezialisierungen homogener und isotroper Turbulenzfelder, schließlich die turbulenten Scherfelder beschrieben, und endlich versucht der Verfasser, die frühen intuitiv gefundenen Lösungen der Göttinger Schule samt ihrer meistens guten experimentellen Bestätigung mit Hilfe des gedanklichen Instrumentariums der statistischen Theorie zu interpretieren. Im ganzen ein sehr gelungenes Buch, wenn man seine Aufgabe darin sieht, das zu bschreiben, was die heutige statistische Turbulenztheorie zu leisten vermag. Das Bewerten der Leistungen dieser Theorie selbst wird aber verschieden ausfallen, je nachdem ob man einen Mathematiker, einen Physiker oder einen Techniker fragt. Der Mathematiker wird bestätigen, daß das geistige Rüstzeug der Wahrscheinlichkeitsrechnung richtig angewandt wurde. Der Physiker wird geltend machen daß zur naturwissenschaftlichen Ergründung des Turbulenzphänomens kein Beitrag geleistet wurde, der Techniker wird die praktisch verwendbaren Informationen für außerordentlich spärlich halten. Hier einige zum Nachdenken anregende Bemerkungen des Buchautors, der im ganzen die Methode der statistischen Theorie durchaus nicht kritisiert, sie sogar für den einzig möglichen Weg hält, zum Lösen aller noch anstehenden Probleme. Man liest: Von der ersten Zielstellung einer Turbulenztheorie, nämlich turbulente Strömungen vorauszuberechnen, ist diese Theorie noch weit entfernt. Durch Anwenden gewisser mathematischer Methoden kann man aus den fundamentalen Differentialgleichungen der Strömungsmechanik neue Gleichungen herleiten, die zwar physikalisch keine zusätzlichen Aussagen beinhalten, die aber bei der Lösung der Gleichungen nützlich sein können und bei der Deutung und Beschreibung der physikalischen Vorgänge — insbesondere in Verbindung mit Versuchsergebnissen — neue Einblicke vermitteln (S. 53). Auch zwischen den weiteren mathematischen Ableitungen liest man eingestreut Feststellungen, die nicht übersehen werden sollten, weil sie doch eine ernste Kritik an der Grundkonzeption darstellen. So auf Seite 56: „Die für Mittelwerte geschriebene REYNOLDSsche Differentialgleichung bildet mit der Kontinuitätsgleichung kein lösbares Gleichungssystem, solange nicht weitere Beziehungen für die REYNOLDSschen Spannungen eingeführt werden." Weiter schreibt der Autor (S. 63): „Das Dilemma, vor das man durch Anwendung statistischer Methoden auf die NAVIER-STOKESschen Gleichungen gestellt wird, besteht darin, den Satz der unendlich vielen gekoppelten Integro-Differentialgleichungen durch weitere Rechenschemata, Hypothesen, Modellvorstellungen und Vereinfachungen oder dgl., die man aus den NAVIER-STOKESschen Gleichungen im allgemeinen nicht erschließen kann, in ein abgeschlossenes Gleichungssystem überzuführen." Näherungsansätze von BOUSSINESQ und PRANDTL bilden eine erste Stufe, aber

„mehr als Aussagen beschränkter Gültigkeit und angenäherten Charakters sind nicht zu erwarten, auch wenn physikalisch begründet erscheinende Schlußfolgerungen aus Daten und Versuchen eingeführt werden."

Während zu Anfang der Bearbeitung dieses Komplexes G. I. Taylor sich noch um das Aufzeichnen von Stromlinienbildern in Wirbelformationen bemühte, ist diese Tendenz heute ganz zurückgetreten. Die aus den Korrelationsfunktionen gefolgerten Längenmaßstäbe sind rein definitorische Maße. Sie sind nicht ohne weiteres als einfache Modelle zur Berechnung der turbulenten Diffusion anwendbar.

Wenn der Autor schreibt: „Die Theorie der turbulenten Strömungen ist ohne Hilfe von Methoden und Vorstellungen der mathematischen Statistik und Wahrscheinlichkeitslehre nicht vorstellbar", dann kann man dem nur unter der Bedingung zustimmen, daß nicht der Anspruch erhoben wird, daß alle Erscheinungen der Turbulenz auf diese Weise speziell mit den am weitesten ausgearbeiteten Methoden für homogene Turbulenz gelöst werden können. Sehr viele turbulente Strömungen in der Entstehungsphase zeigen Strukturbildung und müßten mit ganz anderen Mitteln berechnet werden.

Man muß bedenken, daß eine mathematische Theorie erst dann mit der Aussicht auf Erfolg ausgearbeitet werden kann, wenn eine klare Vorstellung vom physikalischen Ablauf erarbeitet ist, und wenn auch klar gesagt werden kann, was Turbulenz eigentlich bedeutet. Eine Anfangskonzeption, die nur aussagt, Turbulenz sei ein unregelmäßiges Durcheinanderfließen, ist noch kein ausreichendes Fundament, um darauf eine mathematische Theorie aller turbulenten Fließbewegungen aufzubauen. Man muß erkennen, daß es zwar einige wenige Strömungen gibt — im allgemeinen sind es späte Phasen des Abklingens von Wirbelsystemen —, die gut mit den statistischen Mitteln und der Spezialisierung auf Homogenität und Isotropie beschrieben werden können, daß aber diese Mittel nicht ausreichen, alle interessierenden turbulenten Vorgänge zu interpretieren. Aber auch die vergleichbare statistische Theorie der Gase, die das erfolgreiche Vorbild für die entsprechende Turbulenztheorie ergab, ist nicht in der Lage, alle interessierenden Mikrobewegungen zu beschreiben, besonders nicht jene, wo es im ganz Kleinen ebenfalls zu Strukturbildungen kommt, wie das z. B. bei den Kondensationsprozessen der Fall ist.

Dresden W. Albring

K. Jacobs, J. Rosenmüller, Selecta Mathematica IV. (Heidelberger Taschenbücher, Band 98). VIII + 184 S. m. 19 Abb. Berlin/Heidelberg/New York 1972. Springer-Verlag. Preis brosch. DM 14,80.

Die im Rahmen der Heidelberger Taschenbücher geschaffene Reihe ‚Selecta Mathematica' erfährt mit dem vorliegenden Band IV eine wertvolle Fortsetzung. Von den insgesamt fünf Beiträgen sind die ersten vier beliebten Themen des Autors K. Jacobs gewidmet: Grundbegriffe der topologischen Dynamik, Poincarés Wiederkehrsatz, Gleichverteilung mod 1 und Markov-Prozesse mit endlich vielen Zuständen. In gewohnter brillanter Weise versteht es der Autor, bei sicherer Beherrschung mathematischer Präzision den Leser anmutig und fesselnd in tiefliegende mathematische Zusammenhänge einzuführen. Das Herstellen interessanter Querverbindungen und Beziehungen zu Anwendungen sowie anschauliche Interpretation sind dabei weitere hervorhebenswerte Spezialitäten. Diesem Stil wird weitgehend auch der Beitrag von J. Rosenmüller über Konjunkturschwankungen gerecht; er enthält — vielleicht etwas eigenständig — eine gelungene Skizze über Zwei-Personenspiele. Insgesamt vermittelt der Band eine genußreiche und insbesondere auch jüngeren Mathematikern angeratene Lektüre, wenn auch die in der Hauptsache gering gehaltenen Anforderungen an den Leser mitunter ziemlich plötzlich gesteigert werden und auf einige Druckfehler zu achten ist.

Dresden P. H. Müller

H. Bandemer u. a., Optimale Versuchsplanung. (WTB Band 131). 180 S. m. 3 Abb. u. 17 Tab. Berlin 1973. Akademie-Verlag. Preis brosch. 8,— M.

Mit seinem Buch „The Design of Experiments" wies der überragende englische Statistiker R. A. Fisher im Jahre 1935 nachdrücklich drauf hin, daß sich die Aufgabe des angewandten Statistikers nicht auf die Auswertung von Versuchsergebnissen beschränken kann, sondern daß die optimale Planung der Versuche — und zwar im Hinblick auf ihre Auswertung — eine ebenso wichtige Aufgabe darstellt. Etwa seitdem ist die Versuchsplanung Gegenstand statistischer Forschung.

Die Verfasser des vorliegenden Taschenbuchs geben nun eine Einführung in die optimale Versuchsplanung für das Regressionsmodell, wobei sie sich gleichzeitig um eine zusammenfassende Darstellung neuerer Resultate bemühen. Im einzelnen werden drei Themenkreise behandelt, und zwar:

1. die optimale Versuchsplanung im Sinne von J. Kiefer (Kap. 2),
2. die Versuchsplanung — insbesondere für lineare und quadratische Regressionsansätze — im Sinne von G. E. P. Box (Kap. 3) und
3. Versuchspläne zur Modelldiskrimination, d. h. zur Auswahl von Regressionsmodellen (Kap. 4).

Diesen drei Kapiteln wird ein ausführendes Kapitel vorangestellt. Für den Leser dürften außerdem das ausführliche Literaturverzeichnis sowie die Übersetzung wichtiger Fachausdrücke in die russische und englische Sprache von großem Interesse sein.

Zusammenfassend ist festzustellen, daß den Verfassern, die selbst interessante Beiträge zur Theorie der optimalen Versuchplanung leisteten, die Darstellung derzeit bekannter Resultate zur optimalen Versuchsplanung bestens gelungen ist. Mit diesem Buch wurde eine wesentliche Lücke, die seit langer Zeit in der statistischen Literatur bestand, geschlossen. Es ist erfreulich und sicher von großem Nutzen bei der Anwendung statistischer Methoden, daß — wie die Verfasser im Vorwort schreiben — das Risiko der Darstellung eines Wissensgebietes eingegangen wurde, das sich in einer stürmischen Entwicklung befindet.

Dresden V. Nollau

H. Minkowski, Briefe an David Hilbert. 165 S. m. 43 Abb. Berlin/Heidelberg/New York 1973. Springer-Verlag. Preis brosch. DM 32,—.

Nostalgie ist heut modern! So möchte man schließlich die Frage beantworten, zu welchem Ende die von 1885 bis 1908 verfaßten Briefe Hermann Minkowskis an seinen Freund David Hilbert hier veröffentlicht wurden. Die Antwortbriefe sind verlorengegangen. Einen breiten Raum nehmen Hinweise auf Minkowskis zahlentheoretische Forschungen und den geplanten Bericht über die neuere Zahlentheorie ein, für den Historiker eine willkommene Quelle. Zwar ist Geschichte der Zahlentheorie ein seriöses Unternehmen, aber, wie die Dinge einmal liegen, doch wohl von begrenztem Interesse. Größere Aufmerksamkeit würde sich Minkowskis zweites Arbeitsgebiet, Probleme der mathematischen Physik, erfreuen. Doch die Bezugnahme darauf ist vergleichsweise spärlich. Was bleibt übrig? Gewiß, wie die Tochter Minkowskis als Herausgeberin schreibt, „ein klares Bild vom Wachsen der Freundschaft, von der persönlichen Entwicklung der Schreibenden, von dem wachsenden Austausch ihrer Gedanken und der Gemeinsamkeit ihrer fachlichen Interesse". Aber es gibt publizierte Korrespondenzen, — z. B. die zwischen Einstein und Sommerfeld — die so etwas eindrucksvoller belegen, die Umstände der Zeit transparenter werden lassen. Vieles ist allerdings recht witzig. So die Schilderung der Prüfung von fünf Bergbaubeflissenen: „Ich erzielte trotz grenzenloser Unwissenheit der Kandidaten ein glänzendes Resultat, denn die Mehrzahl der Fragen hatte ich so ziemlich auf ja und nein gestellt, und aus dem Tonfall der Frage konnten die Kandidaten einigermaßen entnehmen, ob es angemessener war, ja oder nein zu sagen. Das ganze Collegium von Geheimen Räthen konnte sich nicht vor Staunen fassen, solche bedeutende Kenntnisse waren bei den Kandidaten noch niemals zu Tage getreten." Daneben, was man so über die Kollegen schreibt, viel Berufungsspekulationen, Persönliches. Nun, an eine Veröffentlichung ist natürlich niemals gedacht worden. Aber da sie jetzt vorliegt, geschmückt mit schönen Porträts der wichtigsten zeitgenössischen Mathematiker und einiger Naturforscher, erscheint eine vergangene Epoche in mildem, diffusem Licht. Nostalgie ist heut modern ...

Dresden H.-G. Schöpf

Ed. **E. G. D. Cohen** and **W. Thirring**, The Boltzmann Equation. Theory and Applications. XII + 642 S. m. 1 Porträt u. 85 Fig. Wien/New York 1973. Springer-Verlag. Preis geb. DM 148,—.

Es ist eine schöne Gewohnheit der Wissenschaftler, aus Anlaß von Jahrestagen wegweisender Entdeckungen Symposien zu organisieren, auf denen kritisch gefragt wird: Wie war es? Was ist daraus geworden? Hier nun sind die Vorträge veröffentlicht, die in ehrender Erinnerung an die 1872 erschienene Schrift LUDWIG BOLTZMANNS gehalten wurden, nämlich „Weitere Studien über das Wärmegleichgewicht unter Gasmolekülen." Jene Studien betrafen das Erreichen dieses Gleichgewichts für verdünnte, klassische, monoatomige Gase auf der Grundlage der in dieser Arbeit erstmalig formulierten BOLTZMANNgleichung. Im gewissen Sinne beschreibt sie die wahrscheinlichste oder durchschnittliche Änderung der Einteilchenverteilungsfunktion $f(\boldsymbol{x}\,\boldsymbol{v}, t)$. In der Herleitung war die Beschränkung auf Zweierstöße, die als zufällige Ereignisse behandelt wurden, wesentlich. Wie sich versteht, ergeben sich sofort Fragen, nach der Überwindung all der hier erwähnten Einschränkungen, Fragen, die in diesem Band dann auch behandelt werden, f ist in Wahrheit eine fluktuierende Funktion. Wie also können Fluktuationsterme der BOLTZMANNgleichung hinzugefügt werden? Naheliegender sind folgende Fragen: Wie ist die Voraussetzung der klassischen Mechanik zu überwinden, im Hinblick auf die relativistische und quantenmechanische Verallgemeinerung? Dabei ist letztere weniger akademisch als die erste, denn sie ist spätestens beim Übergang zu polyatomigen Gasen unabweisbar. Wie aber konnten die Ideen BOLTZMANNS über die Theorie der verdünnten Gase im eigentlichen Sinne hinaus entwickelt werden? Wir erwähnen die Transporttheorie der Neutronen und als besonders eindrucksvolle Entwicklung die Anwendung der BOLTZMANNschen (oder einer entsprechenden kinetischen) Gleichung in der Theorie des Festkörpers. Sie ermöglicht sich durch die Konzeption der Quasiteilchen, die, wiewohl keineswegs elementar wie reale Gasteilchen, sich dennoch weitgehend gasartig verhalten. Offenbar erwachsen die größten Schwierigkeiten, wo der Mangel an hinreichender Verdünnung durch diese Konzeption nicht umgangen werden kann, etwa bei realen Gasen. Zunächst wäre nach der Verallgemeinerung der BOLTZMANNgleichung unter Berücksichtigung von Dreierstößen zu fragen und dann generell für Systeme höherer Dichten. BOGOLJUBOW war der erste, der eine solche verallgemeinerte BOLTZMANNgleichung aus der fundamentalen LIOUVILLE-Gleichung gewonnen hat. Aber wie sich nun herausstellte, ist sie mathematisch nicht akzeptabel. Komplementär dazu scheint die Situation in der Ergodentheorie zu sein. Von BOLTZMANN zur Begründung der statistischen Thermodynamik inauguriert, hat sie mathematisch schönste Früchte entwickelt, ohne daß indessen ihre Subtilität in befriedigendem Verhältnis zu dem steht, was sie zur Klärung des eigentlich physikalischen Irreversibilitätsproblems zu leisten vermag. Vielleicht ist hier die Methode der Mastergleichungen erfolgversprechender. Sie steht zwischen BOLTZMANNS Ansatz und dem auf der LIOUVILLE-Gleichung basierenden, insofern wie im letzteren Fall die Verteilung aller N Teilchen in Betracht gezogen wird, und jedoch wie beim ersten Zugang ein stochastisches Element durch die Behandlung der Stöße als Zufallsereignisse auftritt. Es sei aber auf eine höchst interessante Möglichkeit hingewiesen, unter Vermeidung jeder Bezugnahme auf stochastische Prozesse eine Entropie zu konstruieren, welche vermöge der LIOUVILLE-Gleichung zeitlich nicht zunimmt, weil sie nur die Nichtgleichgewichtskorrelationen enthält. Diese Konstruktion basiert auf einer äquivalenten, aber nicht unitär-äquivalenten Darstellung der Dichtematrix, die für die retardierte und avancierte Lösung der LIOUVILLE-Gleichung unterschiedlich ausfällt. Hiermit scheint sich eine Verbindung zwischen zwei fundamentalen, die Zeitrichtung definierenden Phänomenen anzubahnen: thermodynamische Irreversibilität und Kausalität als Bevorzugung der retardierten Lösungen. Wie Verf. (PRIGOGINE) bemerkt, eröffnet sich so eine neue Beziehung zwischen Naturwissenschaft und Philosophie, ganz im Sinne BOLTZMANNS, der tief an philosophischen Problemen interessiert war. Seine wenn auch etwas hausbacken anmutenden, so doch sehr gesunden Ansichten darüber werden im vorliegenden Band ausführlich referiert. Desgleichen finden wir eine sorgfältige Analyse der Entwicklung seiner statistischen Ideen. Eröffnet wird das Buch mit einer Vita aus der Feder seines Enkels, beschlossen mit einigen Anekdoten aus seinen Vorlesungen.

Dresden H.-G. SCHÖPF

F. Klein, Elementarmathematik vom höheren Standpunkte aus. (Die Grundlehren der mathematischen Wissenschaften in Einzeldarstellungen Bd. 14, 15, 16). Band 1: Arithmetik, Algebra, Analysis. XII + 309 S. m. 125 Abb. Berlin 1933. Nachdruck 1968. Preis geb. DM 24,—. Band 2: Geometrie. XII + 302 S. m. 157 Abb. Berlin 1925. Nachdruck 1968. Preis geb. DM 24,—. Band 3: Präzisions- und Approximationsmathematik. X + 238 S. m. 156 Abb. Berlin 1928. Nachdruck 1968. Preis geb. DM 19,80. Verlag von Julius Springer.

Die Vorlesungen über „Elementarmathematik vom höheren Standpunkt aus", die FELIX KLEIN im Jahre 1908 zunächst in autographischer Form herausgebracht hat, gehören zweifellos zu den berühmten Vorlesungen, die zu ihrer Zeit von Göttingen aus auf viele andere Universitäten und Hochschulen ausstrahlten. Sie waren vornehmlich für „Kandidaten des höheren Lehramts" gedacht und hatten das Ziel, den Zustand ändern zu helfen, den F. KLEIN in seinem Vorwort mit den Worten umreißt: „In einer langen Zeitperiode vorher trieb man an den Universitäten ausschließlich hohe Wissenschaft, ohne Rücksicht darauf zu nehmen, was der Schule nottat, und ohne sich überhaupt um die Herstellung einer Verbindung mit der Schulmathematik zu sorgen." Wie weit dieser Zustand heute *echt* überwunden ist, wage ich nicht zu beurteilen; *formal* ist er es sicher dadurch, daß man den Begriff „Schulmathematik" ausgerottet hat und man bemüht ist, einen *inhaltlichen* Gegensatz zwischen der „Mathematik", wie sie als Wissenschaft auf den Universitäten und Hochschulen gepflegt wird und der „Schulmathematik" gar nicht erst aufkommen zu lassen.

Vom Inhalt her vermitteln die Vorlesungen FELIX KLEINS, von denen der Springer-Verlag jetzt einen Nachdruck der vierten Auflage (1933) des ersten Bandes und der dritten Auflage des zweiten Bandes (1925) und des dritten Bandes (1928) herausgebracht hat, indirekt einen tiefen Eindruck von dem Wandel, den die Mathematik in den letzten Jahrzehnten durchgemacht hat, und den Auswirkungen dieser Weiterentwicklung auf die Art und Weise, wie in der Schule Mathematik betrieben wird. Insofern kommt dieser Wiederveröffentlichung große Bedeutung zu.

Band I hat die *Arithmetik* (Das Rechnen mit den natürlichen Zahlen. Die ersten Erweiterungen des Zahlbegriffes. Von den besonderen Eigenschaften der ganzen Zahlen. Die Komplexen Zahlen.), *Algebra* (Reelle Gleichungen mit reellen Unbekannten. Gleichungen im Gebiete komplexer Größen), *Analysis* (Logarithmus und Exponentialfunktion. Die goniometrischen Funktionen. Von der eigentlichen Infinitesimalrechnung) und in einem *Anhang* die Transzendenz von e und π und die Mengenlehre zum Gegenstand.

Band II ist der *Geometrie* gewidmet und behandelt in drei Kapiteln Die einfachsten geometrischen Gebilde, Die geometrischen Transformationen und Die Systematik und Grundlegung der Geometrie. Ein Schlußkapitel gibt einen Überblick über den Unterricht in Geometrie in England, Frankreich, Italien und Deutschland.

Band III beschäftigt sich in drei Kapiteln (Von den Funktionen reeller Veränderlicher und ihrer Darstellung im rechtwinkligen Koordinatensystem; Freie Geometrie ebener Kurven; Von der Versinnbildlichung idealer Gebilde durch Zeichnungen und Modelle) mit der *Präzisions-* und *Approximationsmathematik* und bekundet schon mit dieser Titelwahl, wie eng sich FELIX KLEIN der Problematik der Angewandten Mathematik verbunden fühlte.

Von unschätzbarem Wert — auch für den heutigen Leser — sind die in die Vorlesungen eingestreuten historischen Betrachtungen, die im allgemeinen in der Mathematik viel zu kurz kommen und auch im Unterricht und bei der Ausbildung der Lehrer sträflich vernachlässigt werden.

Dresden H. HEINRICH

EINGEGANGENE BÜCHER

Bei der Schriftleitung sind folgende Bücher eingegangen (ausführliche Besprechung bleibt vorbehalten).

H. Günther, Zur Dynamik schneller Versetzungen. (Forschungsbereiche Kosmische Physik, Heft 3 d. AdW). 50 S. Berlin 1973. Akademie-Verlag. Preis brosch. 10,— M.

L. A. Skornjakow, Elemente der Verbandstheorie. (WTB, Band 130). 177 S. m. 12 Abb. u. 2 Tab. Berlin 1973. Akademie-Verlag. Preis brosch. 8,— M.

N. Elsner, Grundlagen der Technischen Thermodynamik. XI + 660 S. m. 268 Abb. u. 26 Tab., davon 7 als Beilage. Berlin 1973. Akademie-Verlag. Preis geb. 38,— M.

H. Späth, Spline-Algorithmen. (Verfahren der Datenverarbeitung). 134 S. m. 72 Bildern, davon 27 Computer-Programme. München/Wien 1973. R. Oldenbourg Verlag. Preis brosch. DM 40,—.

M. R. Scott, Invariant Imbedding and its Applications to Ordinary Differential Equations. (Applied Mathematics and Computation). XVII + 215 S. m. Fig. London/Amsterdam/Don Mills, Ontario/Sydney/Tokyo 1973). Addison-Wesley Publ. Company. Preis brosch. $ 11.50.

K. Brockhoff, Unternehmensforschung. 121 S. m. 19 Abb. Berlin/New York 1973. Walter de Gruyter. Preis brosch. DM 18,—.

C. Hackl, Schaltwerk- und Automatentheorie II. (Sammlung Göschen, Band 7011). 152 S. m. 117 Abb. Berlin/New York 1973. Walter de Gruyter. Preis brosch. DM 14,80.

D. Greenspan, Discrete Models. (Applied Mathematics and Computation). XIV + 165 S. m. Fig. London/Amsterdam/Don Mills, Ontario/Sydney/Tokyo 1973. Addison-Wesley Publ. Company. Preis brosch. $ 8.50.

G. Busch / H. Schade, Vorlesungen über Festkörperphysik. (Physikalische Reihe, Band 5). 504 S. m. Abb. Basel/Stuttgart 1973. Birkhäuser Verlag. Preis geb. sfr. 68,—.

R. Sigl, Einführung in die Potentialtheorie. (Sammlung Wichmann Neue Folge, Band 4). XII + 232 S. m. 101 Abb. Karlsruhe 1973. Herbert Wichmann Verlag. Preis brosch. DM 38,—.

M. Zyczkowski, Obciazenia złozone wteorii plastycznosci. (Polska Akademie Nauk, Instytut Podstawowych Problemów Techniki). 446 S. m. 149 Abb. (Polnisch, mit englischem Inhaltsverzeichnis). Warszawa 1973. Państwowe Wydawnictwo Naukowe. Preis geb. Zl. 125,—.

W. Görke, Fehlerdiagnose digitaler Schaltungen. (Teubner-Studienskripten, Bd. 60). 230 S. m. 30 Bildern, zahlr. Tafeln u. 45 Beispielen. Stuttgart 1973. B. G. Teubner. Preis brosch. DM 10,80.

L. Collatz / W. Krabs, Approximationstheorie. (Teubner Studienbücher Mathematik). 208 S. m. 39 Abb., 16 Aufg. m. Lösgn. u. zahlr. Beispielen. Stuttgart 1973. B. G. Teubner. Preis brosch. DM 26,80.

K. Stiglat / H. Wippel, Platten. 2. Aufl. VI + 241 S. m. 207 S. Tab., 57 Bilder, div. Tafeln. Berlin/München/Düsseldorf 1973. Verlag Wilhelm Ernst & Sohn. Preis geb. DM 125,—.

G. Márkus, Kreis- und Kreisringplatten unter antimetrischer Belastung. VIII + 189 S. m. 59 Bildern, 75 Tab. Berlin/München/Düsseldorf 1973. Verlag Wilhelm Ernst und Sohn. Preis geb. DM 72,—.

B. v. Querenburg, Mengentheoretische Topologie (Hochschultext) IX + 195 S. Berlin/Heidelberg/New York 1973. Springer-Verlag. Preis brosch. DM 14,80.

NACHRICHTEN

GAMM-Tagung 1974

Die Wissenschaftliche Jahrestagung 1974 der Gesellschaft für Angewandte Mathematik und Mechanik findet in der Zeit vom 1.—5. April 1974 an der Ruhr-Universität in Bochum (BRD) statt. Die örtliche Tagungsleitung liegt in den Händen von Prof. Dr. Ehlich

Conference on Numerical Analysis 1974

The National Committee for Mathematics of the Royal Irish Academy is holding a conference on Numerical Analysis at University College Dublin from 29th July to 2nd August 1974. Invited speakers include:

R. Ansorge (Hamburg)
T. B. Benjamin (Essex)
L. Collatz (Hamburg)
G. Freud (Budapest)
R. Glowinski (Paris)
H. O. Kreiss (Uppsala)
J. D. Lambert (Dundee)
C. Lanczos (Dublin)
P. Rózsa (Budapest)
I. A. Rus (Cluj)
E. Schechter (Cluj)
M. N. Spijker (Leiden)
G. Stampacchia (Pisa)
G. Strang (Cambridge)
F. Stummel (Frankfurt)
V. Thomée (Gothenburg)
J. L. Lions (Paris)
G. I. Marchuk (Novosibirsk)
J. Meinguet (Louvain)
A. R. Mitchell (Dundee)
P. A. Raviart (Paris)
R. S. Varga (Kent)
E. Vitásek (Prague)
A. Wakulicz (Warsaw)
J. H. Wilkinson (Teddington)
M. Zlámal (Brno)

A limited number of short contributed papers will also be accepted. The proceedings of the Conference will be published. Further details and registration forms may be obtained from the Conference Director, Professor J. Miller, School of Mathematics, Trinity College, Dublin 2, Ireland.

Bestellungen sind an das zuständige Postamt oder eine Buchhandlung zu richten, Bestellungen aus dem Ausland an die Firma Buchexport, Volkseigener Außenhandelsbetrieb der DDR, 701 Leipzig, Leninstr. 72, oder direkt an den Akademie-Verlag, DDR-108 Berlin, Leipziger Str. 3—4

Herausgeber und Hauptschriftleiter: Prof. Dr.-Ing. habil. Dr. techn. h.c. H. Heinrich, 8027 Dresden, Friedrich-Hegel-Str. 31. Verlag: Akademie-Verlag, DDR-108 Berlin, Leipziger Straße 3—4; Fernsprecher: 220441, Telex-Nr. 112020, Postscheckkonto: Berlin 35021. Bestellnummer dieses Heftes: 1009/54/2. Die Zeitschrift für Angewandte Mathematik und Mechanik erscheint jährlich mit 12 Heften, einschließlich Tagungsheft. Bezugspreis je Heft 22,— M (Sonderpreis für die DDR 18,— M); Bezugspreis eines Bandes 264,— M zuzüglich Versandkosten (Sonderpreis für die DDR 216,— M). Veröffentlicht unter Lizenznummer 1282 des Presseamtes beim Vorsitzenden des Ministerrates der Deutschen Demokratischen Republik. Gesamtherstellung: VEB Druckerei „Thomas Müntzer", 582 Bad Langensalza.
Printed in the German Democratic Republic.

JOHANNES THOMAS

Mathematische Theorie der Aerodynamik des Skifluges

(Schriftenreihe der Institute für Mathematik der Akademie der Wissenschaften der DDR, Reihe B)

1971. 92 Seiten – 9 Abbildungen – gr. 8° – 20,– M
Bestell-Nr. 761 4421 (2069/B/11)

Gegenstand der Monografie ist die analytische Ermittlung der möglichen Flugbahnen eines Skispringers nach dessen Absprung vom Schanzentisch einer Sprungschanze. Die Ermittlung dieser Flugbahnen erfolgt auf der Grundlage allgemeiner aerodynamischer Gesetze nach der hierfür von dem Numeriker Herman König eingeführten Hodografenmethode für den Fall, daß an der Schanze kein Hangwind weht. Die Hodografengleichung wird durch eine Transformation in die Gestalt einer Abelschen Differentialgleichung zweiter Art mit rationalen Koeffizientenfunktionen gebracht, von der gezeigt wird, daß sie nicht durch Quadraturen, wohl aber zeichnerisch integriert werden kann. Nach Betrachtungen über den globalen Verlauf der Lösungskurven der Abelschen Differentialgleichung werden diese Lösungskurven analytisch durch Potenzreihen beschrieben, wobei hinsichtlich Geschwindigkeit und Richtung des vom Schanzentisch abspringenden Skispringers mehrere Fälle unterschieden werden müssen. Für jede der genannten Potenzreihen werden eine Majorante und ein Konvergenzbereich angegeben. Der Fall des Vorhandenseins von Hangwind an der Schanze wird an den hangwindfreien Fall durch Störungsrechnung angeschlossen.

Die in der Arbeit gemachten Betrachtungen sind deduktiv-mathematischer Natur. Ihre praktische Bedeutung besteht darin, auf ihrer Grundlage Anhaltspunkte für Sprungschanzenkonstruktionen zu erhalten.

Bestellungen durch eine Buchhandlung erbeten

AKADEMIE-VERLAG
DDR-108 Berlin, Leipziger Str. 3–4

MIX
Papier | Fördert gute Waldnutzung
FSC® C083411

Zeitfracht Medien GmbH
Ferdinand-Jühlke-Straße 7
99095 Erfurt, Deutschland
produktsicherheit@kolibri360.de